Plant Biotechnology and Molecular Biology

A Laboratory Manual

By

M. S. Punia

Department of Plant Breeding
CCS Haryana Agricultural University
Hisar -125 004 (India)

SCIENTIFIC PUBLISHERS
Research · Publish · Inspire

Published by
SCIENTIFIC PUBLISHERS (INDIA)
5 A, New Pali Road, P.O. Box 91
Jodhpur 342 001 (India)
E-mail: info@scientificpub.com
Website: www.scientificpub.com

Print : 2019

ISBN : 978-81-72332-03-7
E-ISBN : 978-93-88172-36-3

Visit the Scientific Publishers (India) website at
http://www.scientificpub.com
Printed in India

Preface

Biotechnology and molecular biology have made significant progress and perhaps one of the most exciting areas of modern science is their application to the study of plant system and genetic improvement of crop plants. They offered new tools for opening up new research avenues to develop novel genotypes, organisms, biological systems and commercial products, with unlimited potential to resolve varieties of problems related to agriculture, industry, environment, health, energy and other important areas. The application of genetic engineering or recombinant DNA technology to crop improvement programmes has resulted in breaking all the barriers of conventional gene transfer by plant breeding approaches which allow incorporation of genes from only closely related species, whereas, it is possible to transfer and clone gene(s) from organism to the plants by recombinant DNA technology.

The book, "A Laboratory Manual of Plant Biotechnology and Molecular Biology" comprises of workable laboratory protocols for a large number of techniques related to plant biotechnology, genetic engineering and molecular biology. This includes plant cell and tissue culture, callus and suspension culture, anther culture, ovule culture, embryo culture, Cryopreservation, Isolation of Plant protoplasts, Protoplast culture and regeneration, production of somatic hybrids through protoplast fusion, gene transformation using Agrobacterium as vector, direct gene transfer using biolistic gun, Isolation of plant and organells DNA, construction and screening of genomic DNA libraries, Molecular markers like RFLP, RAPD, SCARS and CAPS, DNA sequencing, RNA isolation and northern blotting, Isolation of proteins and western blotting etc.

The manual is prepared with the objective to cater the needs of post- graduate students as well as for scientists working in the disciplines of Plant Breeding, Genetics, Botany, Plant physiology, Biochemistry, Plant Biotechnology, Molecular Biology etc. It gives an update on some well established methods and presents reliable protocols. I hope that the users will find this manual as useful tool for the accomplishment of their research projects.

Place : **Hisar** (Haryana)

Date : August, 1998. **M. S. Punia**

Contents

APPENDICES

1

ORGANIZATION OF PLANT TISSUE CULTURE LABORATORY

A plant tissue culture laboratory must have adequate space for performance of several functions. The space for housing different facilities required for plant tissue culture varies from laboratory to laboratory. For each basic facility well cleaned separate spacious rooms are required for housing equipments and to ensure proper working conditions. There are a number of basic facilities and a minimum level of organization which should be available to either the individual or to a team.

The basic facilities for a tissue culture laboratory usually include the following :

A. Wash-up area.

B. Media preparation area/room

C. Culture inoculation and transfer area.

D. Culture incubation room/growth room.

E. Observation/Experimental area

F. Green house/glass house.

A. WASH-UP AREA

An area with large sink (preferably stainless steel), draining bench top/racks is necessary. This area may have sufficient space for the following :

(i) Drying oven.

(ii) Baths for acid and detergent treatments of glasswares.

(iii) Baths for pipette washing.

Ready access to demineralized, distilled and double distilled water is necessary.

A sufficient amount of good quality detergent/surf/teepol should always be available for washing glassware. The cleaned glassware is inspected and dried at 150°C in a convection drying oven for at least 1 hr. capped with aluminium foil and stored in a closed cup board to limit dust collection. Highly contaminated glassware and pippettes are routinely first treated in chromic bath for at least overnight or go for ultrasonic cleaning.

Glassware : Tissue culture requires the same glasswares as required in a well stocked laboratory. Beyond the normal range of glasswares, wide necked Erlenmeyer flasks (50-250 ml.), culture tubes (25×150-250 mm), and glass Petri dishes (60×15 and 100×20 mm) are useful as tissue culture containers. However, culture tubes (to reduce the space requirement) with polypropylene caps (to reduce water evaporation) are widely used culture vessels for tissue culture. Beyond the normal range of glassware and instruments found in a well stocked laboratory, a plentiful supply of graduated measuring cylinders (10-1000 ml), volumetric flasks (100-3000 ml) and culture bottles, wide mouth gem bottles (250 ml) are particularly useful for tissue culture experiments.

Plastic Labware: Presterilized disposable plastic Petri dishes reduce the breakage and amount of washing. Due to its high cost, it may not be useful for routine tissue culture work. Reusable plastic labware comes in many different types, and their chemical and heat resistance vary tremendously. It is recomended that plastic labwares should be washed with mild, nonabresive detergent, followed by rinse with tap water and then distilled water.

B. MEDIA PREPARATION AREA

This area preferable a room, should have sufficient storage and bench space for chemicals, glasswares, culture vessels and other items needed for media preparation. The media room may require the following equipments.

(a) Refrigerator for safe storage of chemicals, media, stocks solutions etc.

(b) Distilled water plant.

(c) Hot plate and magnetic stirrer.

(d) pH meter.

(e) Precision balance (-to 5 decimal place).

(f) Simple balance (-to 2 decimal place).

(g) Bunsen burner with a gas source.
(h) Metal or wood, racks to support culture tubes.
(i) Metal trolleys for transport of culture racks, flasks, tubes etc.
(j) An oven.
(k) Microwave oven.
(1) Autoclave.
(m) Culture vessels/test tubes.
(n) Millipore filters of different sizes.
(o) All the chemicals used for media preparation should be of analytical grade.
(p) Automatic micropipettes/glass pipettes of different volumes.

Water : To obtain high purity laboratory-grade water, glass distillation unit is usually used. However, newly introduced reverse osmisis purifying equipment (e.g. Milli-RO Reverse Osmosis system) can be installed depending on the amount of water needed because this equipment can deliver highly pure water at a rate of 4, 15 or 40 liters per hour depending on the model.

Sterilization area/autoclave room : The widely used method of sterilizing equipments, media, water, solutions etc. is by autoclaving at 121°C with a pressure of 15 psi for 15 minutes. A modern autoclave with pressure gauge, slow exhaust is essential for rigorous exclusion of contaminating microorganisms. Bacteria proof membrane filters (Millipore^R by Millipore Corp) fitted with 1-10 ml graduate syriges by a Swinny^R filter adapter or Millipore^R vacuum filtering set-up is necessary for filter sterilization of heat labile compounds like some growth regulators.

C. CULTURE INOCULATION/TRANSFER AREA

The most commonly used accessory for aseptic culture inoculation or transfer is the laminar air-flow cabinet. However, the laminar flow cabinet should be installed in a spaceous but clean room preferably with controlled temperature. The various miscellaneous instruments/ equipments required for aseptic culture inoculation include-forceps, scalpel, spirit lamp/bunsen burner. A simple dissecting stereomicroscope is necessary for meristem isolation in case of meristem culture work.

D. CULTURE INCUBATION ROOM/GROWTH ROOM

Incubation of cultures under controlled temperature, light and humidity regimes is essential for the proper growth and phenotypic expression of a culture. To satisfy this, a culture room should be provided with the following :

1. Well insulated room preferably away from direct heat, dust and air flow.
2. Provided with cooling and heating device to control the temperature in the range of 20-30°C.
3. Should be equipped with shelves (preferably 3 tiers) fitted with fluorescent tubes to provide light intensity on each shelf from 1000 to 6000 Lux.
4. Lights should be connected to a timer for the control of photoperiod.
5. Temperature recorder.
6. Thermostat controlled cut-out device for the whole illuminating system of the room.
7. Stand-by generator for emergency power supply.

E. OBSERVATION AREA

An area or room adjacent to the culture room provided with microscopes particularly a dissecting stereozoom microscope is necessary for periodical observation of growth and development of cultures.

F. GREEN HOUSE/GLASS HOUSE FACILITY

A greenhouse/Glasshouse facility is essential for pre-inoculation planting materials. It is also required for preliminary transfer of tissue culture regenerated plants to *in vivo* growth. The size of the greenhouse depends on the individual laboratory requirement. Sometimes a glasshouse with controlled temperature, light and humidity may be required for growing plants in off season.

REFERENCES

1. Biondi, S. and Thorpe, T.A. 1981. Requirements for a tissue culture facility. *In :* (Ed. Thorpe, T.A.) Plant Tissue Culture : Methods and Applications in Agriculture. Academic Press, New York. pp. 1-20
2. Brown, D.C.W. and Thorpe, T.A. 1984 Organization of a plant tissue culture laboratory. In : Cell Culture and Somatic Cell Genetics of Plants., (Ed. Vasil, I.K.) Academic Press, New York. Vol. 1 : 1-12.

2

Media Preparation, Sterilization and Storage

PREPARATION OF PLANT CELL AND TISSUE CULTURE MEDIA

The most efficient way of preparing plant cell culture media is first to make up stock solutions of major inorganic nutrients, trace elements, iron source, vitamins, and individual plant growth regulators. An example of how to do this for the MS (Murashige and Skoog, 1962) medium is given in Table 1. Store the vitamins frozen at -20°C in small batches, thaw, and mix fully before use. The other stocks can be kept in a refrigerator at 4°C, but should be frequently checked and discarded if precipitation occurs; do not store inorganic stock solutions longer than one month. It is advisable to make up small stocks of growth regulators fresh for each batch of media, as small changes in concentration due to precipitation can seriously affect growth of the cultures.

For semi-solid media, add agar at a final concentration of 6-10 g/litre prior to autoclaving. It is important to use a good quality, bacteriological grade agar for plant cell culture work. For pouring of agar medium in Petri dishes, it is convenient to autoclave 500 ml batches in 1 litre conical flasks to dissolve the agar, allow the medium to cool to around 40°C, and then pour.

Note that, although the pH of most plant culture media is adjusted to pH 5.8 prior to autoclaving, the final pH may be slightly lower due to the production of sugar acids during autoclaving.

The length of time required to sterilize a batch of culture medium by autoclaving will depend upon the volume of medium in the flask. For 250 ml batches, 25 minutes at 121°C is adequate. Thermolabile growth regulators (see appendix 2) must be added to the medium after it has cooled to 40°C (semi-solid media) or below.

Composition of commonly used plant tissue culture media :

Composition of media can be divided into six groups, these are :

1. Major inorganic nutrients.
2. Trace elements.
3. Iron source.
4. Vitamins.
5. Carbon Source.
6. Plant growth regulators.

These divisions are both functional and practical i.e. they reflect the way in which the various stock solution are made up as given in Table 1.

Protocol

1. Prepare stock solutions of major inorganic nutrients, trace elements, vitamins, and plant growth regulators using analytical reagent grade chemicals and distilled water. For protoplasts media, use double distilled water.
2. For 1 litre of liquid medium, pipette the required volumes of each stock solution into 1 litre glass beaker on a magnetic stirrer.
3. Add sucrose and agar as solid, Stir until it is fully dissolved, add more water if necessary.
4. Adjust the volume to about 950 ml with distilled water.
5. Adjust the pH to the required value (usually 5.8) with 1N NaOH and 1N HCl.
6. Transfer to a 1 litre measuring cylinder or volumetric flask, and adjust volume to 1 litre with distilled water.
7. Transfer back to the flask and stir for complete mixing.
8. Transfer 75 ml batches of medium to clean 250 ml conical flasks or 10-15 ml in test tube plug with non-absorbent cotton wool or sponge plugs, cover the tops with aluminium foil, and autoclave for 15 min. at 121°C and 15 psi.

Note 1. Store inorganic stock solutions at 4°C for a maximum of one month. Dispense vitamins in 10 ml. aliquots and store at -20°C. Prepare growth regulator stocks fresh each time; auxins are usually titrated into solution with NaOH, whereas, cytokinins are dissolved in dilute NaOH or aqueous ethanol. If growth regulators are thermolabile, add them filter sterilized after the autoclaved medium has cooled.

'Concentrated media may be stored frozen (-20°C) at this stage. Store the sterilized medium at 25 to 30°C for its subsequent uses.

2. The composition of various media viz. MS, SH, B5, KM, LM, MN, and SI is given in appendix 1.

Table 1 : Composition and Preparation of Murashiage and Skoog Medium (MSO)

Stock	Component	Mg/l	Concentration	Volume of stock per litre of medium (ml.)	Storage of stock solutions
Stock A	$CaCl_2$ $2H_2O$	44000	× 100	10	+4°C
Stock B	NH_4 NO_3	16500	× 10	100	+4°C
	KNO_3	19000			
Stock C	KI	83	× 100	10	+4°C
	$CoCl_2$ $6H_2O$	2.5			
Stock D	KH_2 PO_4	17000	× 100	10	+4°C
	H_3 BO_3	620			
	Na_2, MoO_4 $2H_2O$	25			
Stock E	$MgSO_4$ $7H_2O$	37000	× 100	10	+4°C
	$MnSO_4$ $4H_2O$	1700			
	$CuSO_4$ $5H_2O$	2.5			
	$ZnSO_4$ $7H_2O$	860			
Stock F	$FeSO_4$ $7H_2O$	5570	× 200	5	+4°C
	Na_2 EDTA	7450			
MS	Myo Inositol	10000	× 100	10	-20°C
Vitamins	Nicotinic acid	50			
	Pyridoxin HCl	50			
	Thiamine HCl	50			
	Glycin	200			
	Sucrose	30000			
	Agar	7000			
	pH	5.8			

REFERENCES

1. Dixon, R.A. (ed.) (1985). Plant Cell Culture : A Practical Approach, IRL Press, Oxford, Washington DC.
2. Evans, D.A., Sharp, W.R. Ammirato, P.V., and Yamada, Y. (ed.) (1983). Handbook of Plant Cell Culture, Vol. 1, Macmillan Publishing Co., New York.
3. Gamborg, O., Miller, R., and Ojima, K. (1968). Exp. Cell Res., 50, 148.
4. Gresshoff, P.M. and Doy, C.H. (1974). Planta, 107, 161.
5. Kao, K.N. and Michayluk, M.R. (1975). Planta, 126, 105.
6. Litvay, J.D., Verma, D.C. and Johnson, M.A. (1985). Plant Cell Rep., 4, 325.
7. Franklin, C.I., Trieu, T.N. Gonzales, R.A., and Dixon, R.A. (1991). Plant Cell Tiss. Organ Cult, 24, 199.
8. Murashige, T. and Skoog, F. (1962), Physiol. Plant., 15, 473.
9. Nitsch, J.P. and Nitsch, C. (1969). Science, 163, 85.
10. Schenk, R.U. and Hildebrandt, A.C. (1972). Can. J. Bot., 50, 199.
11. Vasil, I.K. (ed.) (1984). Cell Culture and Somatic Cell Genetics of Plants, Vol. 1, Academic Press, New York.

3

ASEPTIC SEEDLINGS PRODUCTION AND STERILIZATION OF *IN VIVO* GROWN PLANT MATERIALS AND EXCISED EXPLANTS

A. PRODUCTION OF ASEPTIC SEEDLINGS *IN VITRO*

Aseptic seedlings are produced following the below given protocol :

1. Take healthy seeds and remove their seed coat (for dicots) if seed coat is hard.
2. Put all the seeds in a small piece of musline cloth and tied its top with a small thread.
3. Put the seeds in a 0.1% $HgCl_2$ or $NaoCl_2$ solution for 5 to 15 minutes depending upon the species, add 2-3 drops of tween-20.
4. Remove the seeds from $HgCl_2$ solution and give 7-8 washes of sterilized distilled water in a laminar airflow box.
5. Plating of the seeds in done in the MSO basal medium in the test tubes.
6. Incubate it at 25°C ± 1°C temp & 16/8 hrs light/dark cycles of photoperiod.
7. The seeds will germinate in due couse of time and produce aseptic seedlings.

These seedlings are used for excising the explants for further tissue culture experiments.

B. STERILIZING *IN VIVO* GROWN PLANT MATERIAL

To avoid bacterial and fungal growth, which is detrimental to culture growth, explants are surface sterilized before they are used to establish *in vitro* axenic cultures. The most common disinfectants are listed below with their concentration and exposure times that preserve the explant but at the same time destroy any microbial contamination.

PROCEDURE

1. Wash explants in a mild detergent before treatment with the disinfesting solution. (Herbaceous material may not require this step)
2. Rinse explants throughly under running tap water for 10-30 minutes.
3. Submerge explants into the disinfestant solution. Seal bottle and gently agitate.
4. Under sterile conditions, decant the solution and rinse explants several times with sterile distilled water, before plating on cultural media.

Table 2 : Commonly used Disinfestants for Plant Tissue Culture

Disinfestant	Concentration (%)	Exposure Time (min.)
Calcium hypochlorite	9-10	5-30
Sodium hypochlorite[a]	0.5-5	5-30
Hydrogen peroxide	3-12	5-15
Ethyl alcohol	70-95	0.1-5.0
Silver nitrate	1	5-30
Mercuric chloride	0.1-1.0	2-10
Benzakonium chloride	0.01-0.01	5-20

[a] Commercial bleach contains about 5% sodium hypochlorite and thus may be used at a concentration of 10-20% which is equivalent to 0.5-10% sodium hypochlorite. Sterilization procedures may be enhanced by :

1. Placing the material in a 70% ethyl alcohol solution prior to treatment with another disinfectant solution. The use of a two step (two source) sterilization procedure has proven beneficial with certain species.
2. Using a wetting agent such as Tween 20 or 80 can be added to the disinfectants to reduce surface tension and allow better surface contact.
3. Conducting the sterilization process under vacuum. This result in the removal of air bubbles and provides a more efficient sterilization process.

C. STERILIZATION OF EXPLANTS EXCISED FROM FIELD *(IN VIVO)* GROWN PLANTS

Protocols

1. Cut the twigs, branches or inflorescence, from field grown nurseries, put them into a jar containing fresh water for their transportation to laboratory.
2. Cut the explants to a small size and wash with soap solution.
3. Wash gently in the running tap water.
4. Wrap the explants in musline cloth and dip them in a ethyl alcohol for 10-30 seconds.
5. Put the explants in 0.1% $HgCl_2$/$NaoCl_2$ solution adding 2-3 drops of Tween-20 for 5-10 minutes.
6. Take out the explants from $HgCl_2$ solution and put them into sterilized sieve and give 7-8 washing with sterilized distilled water. Transfer to the sterilized petri dish.
7. Cut the explants to an appropriate sizes with sterilized scalpel & forceps before their plating on the medium.

Note : Steps 5-8 should be carried out in the laminar flow under aseptic condition.

4

MICROPROPAGATION

Micropropagation refers to quicker and faster multiplication of plant propagating material *in vitro* for raising true to the type disease free seedlings/nurseries in a limited time and space. A wide range of plants have now been regenerated and propagated through the techniques of cell, tissues and organ culture (Fig. 4.1) There are three major benefits of this method : (i) rapid multiplication of superior clones and maintenance of uniformity, (ii) multiplication of disease free plants and (iii) multiplication of sexually derived sterile hybrids.

PLANT MATERIAL FOR MICROPROPAGATION

The starting material for micropropagation is a shoot tip or auxiliary buds.

Stages of micropropagation :

1. Selection and sterilization of elite plants
2. Establishment of auxiliary buds/shoot tip in culture.
3. Multiplication (proliferation) in culture.
4. Rooting of *in vitro* plants and transfer to compost / soil.

Protocols

1. Shoot tip or auxiliary buds (young growing) should be excised as an explants from field grown plants.
2. Wash and disinfect it with soap solution and wash it with distilled water. Put the explants in 0.1% $HgCl_2$ for 5-7 min. containing 3-4 drops of Tween 20. Give 5-6 washing with sterilized distilled water in laminarflow and put the explants in sterilized petri dishes.
3. After surface sterilization, explants are further trimmed of f to an appropriate sizes and place them on suitable establishment medium. The establishment medium varies from species to species.

4. Incubate the culture at 25 ± 1°C under 16/8 hrs. light/dark cycles of photoperiod.
5. This is the phase for acclimatization of the tissue to the *in vitro* conditions which requires about 3-4 weeks and varies from species to species.
6. Under optimum laboratory conditions, the small shoot tips/auxiliary buds grow well in a normal fashion, giving rise to small, unrooted shoots in culture.
7. The initiated shoots should be subcultured so that the nodel section of the shoot are placed onto multiplication (proliferation) medium which contains high concentration of cytokinins which are likely to give rise to precocious shooting of the auxiliary buds.
8. The normal period between subculture is 3 weeks and during this time a number of shoot tips are available for subculture will multiply as a result of breaking of nodal auxiliary buds anywhere from 2 to 12 times.
9. After 3 weeks on multiplication (proliferation) medium on an average of about 5 shoot tips are obtained which subsequently subcultured on fresh medium separately and individually. These inturn will multiply 5 times in the subsequent 3 weeks and thus micropropagation has led to a 25 fold increase to plant material available over a total of about 6-8 weeks.
10. Rooting of the above micropropagated shoots are carried out normally is MS medium supplemented with a low concentration of IAA (1 mg/l).
11. Once the micropropagated plants are rooted satisfactorily, they are transferred to soil/soil rite/soil substitutes, they must be weaned (hardened) to a normal humidity level.
12. During the process of establishment in soil, they require about 100% humidity, which inhibits the normal development of cuticle and stomatal activity associated with drier conditions. Normally it requires reduced lighting, therefore, plants must not be exposed for the first time to full sunlight. This in done slowly for a period of couple of weeks.
13. This procedure will develop small, autonomous and photosynthesizing plants. During hardening phase for rooting in soil under reduced lighting conditions, they live on carbohydrate reserves build up during micropropagation period *in vitro*.

14. Once this weaning (hardening) process is completed, the hardened off plants can be incorporated into normal growing system for their further development.

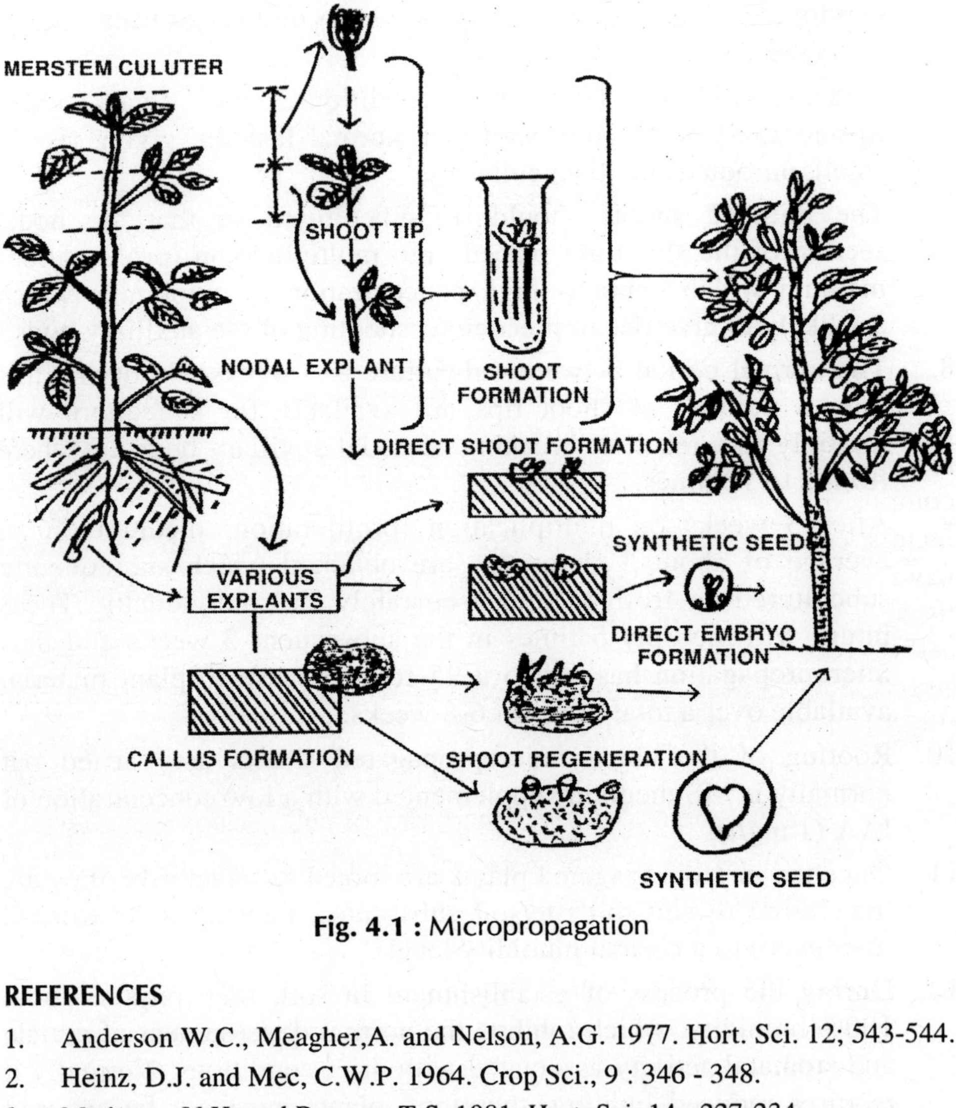

Fig. 4.1 : Micropropagation

REFERENCES

1. Anderson W.C., Meagher,A. and Nelson, A.G. 1977. Hort. Sci. 12; 543-544.
2. Heinz, D.J. and Mec, C.W.P. 1964. Crop Sci., 9 : 346 - 348.
3. Mathews, V.H. and Rangan, T.S. 1981. Hort. Sci. 14 : 227-234.
4. Mathews, V.H., Rangan, T.S. and Narayanaswami, S. 1976. Z. Planzenphysiol. 79 : 450-454.
5. Shah, R.R. and Dalal. K.C. 1980 Curr. Sci., 49 : 69-71.
6. Sharma, A.K. and Mitra, G.C. 1976. Indian J.Exp. Biol., 14 : 348-350.

5

INITIATION AND MAINTENANCE OF CALLUS AND SUSPENSION CULTURES

Callus is an unorganized proliferative mass of plants cells. The general growth characteristics of a callus involves a complex relationship between the plant material used to initiate the callus, the composition of medium, and the environmental conditions during the incubation period (Dodds and Robert, 1985). Callus can be induced from a veriety of explants / tissues. Callus formation from explant tissue involves the development of progressively more random plans of cell division, less frequent specialization of cells and loss of organised structures (Thorpe, 1980). There are five phases of callus growth (Phillips and Dodds 1977).

1. A lag phase, where cells prepare to divide.
2. An exponential phase, where rate of cell divison is highest.
3. A linear phase, where cell division slows but the rate of cell expansion increases.
4. A deacceleration phase, where the rate of cell division and elongation decrease.
5. A stationary phase, where the number and size of cell remain costant.

Callus and cell cultures having the potential to produce plantlets via somatic embroyogenesis are referred to as embryogenic cultures. Choosing the right ex-plant is very critical for establishing embronic cell cultures.

Protocols for callus induction from sunflower explants (Punia and Bohorova, 1992)

1. Dehusk the sunflower seeds with fine forceps

2. Sterilized the seeds in 0.1% $HgCl_2$ for 15-20 minutes.
3. Give 6-8 washing with sterilized distilled water and plate 2 seeds in each test tube containing MS_0 medium for raising aseptic seedlings in growth room having temperature 25 + 1°C & 168 hrs photoperiod.
4. Explants viz. hypocotyl segments, leaf segments and cotyledons segments are excised from 7-11 days old grown aseptic seedlings.
5. Explants are plated on MSD4 (MS + 0.5 mg/l BAP + 0.05 mg/l NAA) medium in a wide mouth conical flaskes.
6. Transfer the cultures flaskes to the growth room and incubate at 25 + 1°C, 16/8 hrs. photoperiod.
7. Callus formation starts 4-5 days after plating the explants, Subculture the calli after every 20-25 days to produce compact greenish large masses of calli (Fig 5.1).
8. To regenerate plants, transfer well developed calli on solidified R (MS +1.0 mg./l BAP + 0.1 mg/L GA_3 + 500 mg/l casamino acid + 40 ml/L adenine sulphate) medium which differentiate the calli into small emeryoids and subsequently into plant lets (Fig. 5.2).

Fig. 5.1. Callus Development

SUSPENSION CULTURE

Cell suspensions are initiated by transfer of undifferentiated callus fragments/pieces into flasks with a liquid medium, which are then placed on a gyratory shaker (125 rpm) to provide aeration to the cells. A large inoculum may be necessary to initiate the suspension and agitating rates on orbital shakers should be in the range of 30-150 r.p.m. with an orbital motion stroke of clumps and the necessary inoculum size of freely suspended cells and small cell clusters may not be attained. When the newly established suspension cultures has reached a suitable cell density for sub culture, remove as much of the remaining unbroken callus material and large clumps as possible. This can be done by either transferring the single cells and small clumps with a sterile syringe, by filteration, by allowing the large material to settle down and pipetting off from the top of the culture or simply by pouring. An optimum density of cells in suspension culture is must for their normal growth which depends upon the rate of growth and composition of the media. Generally, ten percent of initial cell density to total volume of the cultures will be sufficient. Cell suspensions can be maintained by subculturing in flasks, called batch cultures. Cells from suspension culture can be harvested by filteration or transferred to agar solidified medium and grown as callus for its further differentation into plantlets on regeneration medium.

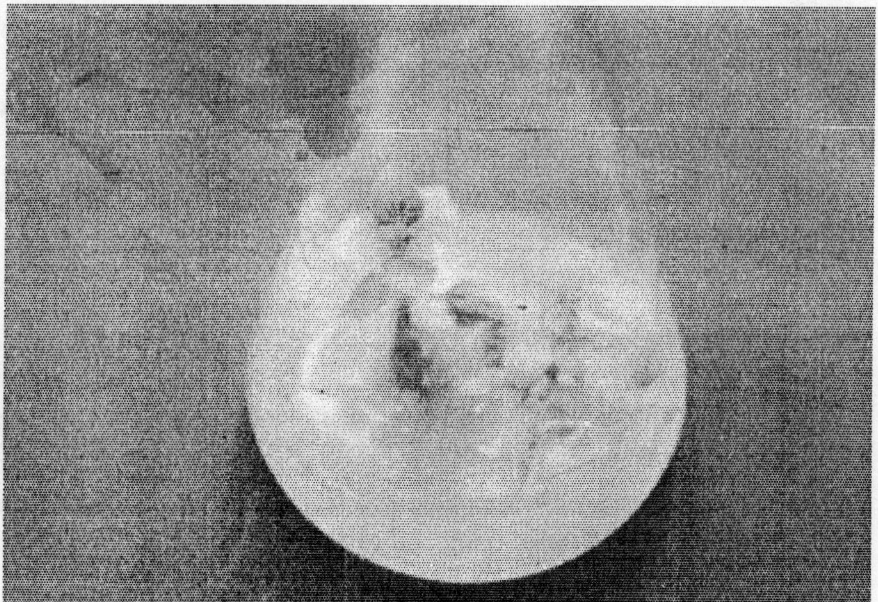

Fig. 5.2. Plant Regeneration

A. INITIATION OF SUNFLOWER SUSPENSION CULTURE AND REGENERATION

Protocols

1. Prepare MSP_l (MS + 0.5 mg/l NAA+ 0.1 mg/l BAP)liquid medium (without agar), pour 25 ml /flash & sterilize it into autolave.
2. Pick up well developed calli of sunflower (See protocols for callus development), cut them into small fragments & put 500-750 mg of callus into the liquid medium in a limnar hood.
3. Keep the flaskes on the shaker and agitate them continuously at 80-100 rpm for about 2 weeks in culture room at a temperature of 25 + 1°C.
4. Filter the medium on a sterlized watman filter paper.
5 Plate the suspension harvested cells on agar solidified MSD_4 medium for callus development.
6. Transfer well developed undifferentiated calli on R medium for differentiation into plantlets.
7. Allow plantlets to grow inside the flask.
8. Separate the plantlets from each other and place them into MS_0 medium for rooting.
9. Hardened off the fully developed rooted plants and transfer them to pots containing soil rite mixture into green house.
10. At flowering, pollinate the flower continuously for 5-7 days & harvest the seeds.

B. ESTABLISHMENT OF EMBROYOGENIC SUSPENSION CULTURE OF RICE *(ORYZA SATIVA)* AND PLANT REGENERATION

Protocol

a. Establishment of embryogenic cultures

1. Surface sterilize mature, dehusked rice seeds by treating with bleach (5.25% $NaHClO_4$) for 30 min. under constant agitation.
2. Rinse seeds three times with sterile distilled water.
3. Repeat steps 1 and 2 two more times to ensure disinfestation (mature seeds are able to tolerate harsh disinfestation treatments compated to other plant materials).
4. Plate seeds embryo side up on N6 medium (Hooker and Nabors, 1977) containing 12μM proline, 20μM 2,4-D (4.4

mg/litre) (N6-20 medium), and 0.25% Gelrite.

5. After three days, transfer the embryos that are not contaminated to fresh N6-20 medium.
6. After eight days, dissect the scutellar callus dome from the endosperm, root, and shoot. If non-embryogenic callus is present on the callus dome, remove and discard.
7. Transfer the embryogenic callus to solid N6 medium for two to four weeks.
8. For initiation of embyogenic suspension cultures, inoculate 0.5 - 1.0 g fresh weight of pure embryogenic callus into 50 ml. of liquid N6 medium containing 12 mM proline, 20μM (4.4 mg/1) 2,4-D and 0.34 mM glutamine.
9. Examine the suspension cultures for quality. If necessary, perform selective subculture on the suspension culture material.

b. Regeneration of plants

1. For embryo development, collect the suspension culture material and rinse with liquid growth regulator- free N6 medium. Plate tissue on solid N6 medium containing 12mM proline, 50μM ABA, 6% socrose, and 2-5μM (0.4-1.1. mg/l) 2,4-D, and place in the dark for two weeks.
2. Transfer the callus to N6 medium containing 12mM proline and 6% sucrose and place back in darkness for two to four weeks. Examine the cultures periodically for the appearance of somatic embryos. If few or no somatic embryos appear after four weeks, repeat steps 8 and 9.
3. Remove mature somatic embryos and plate on MS medium containing 100 mg/litre inositol, 5 mg/litre thiamine HCI, 2g/litre casein hydrolysate, 2.3μM kinetin, 8g/litre agar for germination.
4. Transfer plantlets continuing shoots and roots to soil after they reach a height of 10 cm. Grow initially in the shade with moderate misting.

Table 1. Media and growth regulators levels used for callus and cell cultures of selected species.

Species	**Explant**	**Medium**	**Growth regulators (µM)**	**Reference**
Allium cepa	Stem	S	2,4-D 1.3	Donstan and Short (1977)
Arabidopsis thaliana	Seeds, stem, or leaf	S	2,4-D4.5-9 + K2.3	Negrutiu *et. al.* (1975)
Arachis hypogea	Cotyledon	MS	Pic 2.1	Ozias-Akins *et. al.* (1992)
Asparagus officinalis	Hypocotyl	LS	2,4-D 4.5 + K 1.5	Wilmar and Hellend (1968)
A vena sativa	Leaf or tiller buds	MS	2,4-D 9 + BAP 4.4 + CM 10%	Shekhawat *et. al.* (1984)
Aylostera pseudodeminuate	Vascular Tissue	S	2,4-D 5 K 5.1	Havel and Kolar (1983)
Beta vulgaris	Hypocotyl	S	NAA 53.7 + K 4.7	Hooker and Nabors (1977)
Catharanthus roseus	Leaf	B5	2,4-D9	Stafford *et. al.* (1985)
Cattleya sp.	Meristem	S	NAA 0.5 + K 0.9 + CM 15%	Lindemann *et. al.* (1970)
Cinchona ledgeriana	Hypocotyl	B5	2,4-D 4.5+ K 0.9	Hanker *et. al.* (1985)
Cofea arabica	Leaf	S	2,4-D + K2	Sondahl *et. al.* (1984)
Crassula argentea	Leaf	MS	Zea 108.2 (or) 2iP 98.4	Paterson and Rost (1979)
Cucumis sativus	Internode	MS	NAA 1 + BAP 0.1	Aziz and McCown (1985)
Cymbidium sp.	Shoot apex	Wh	NAA 26.9 + CM 10%	Steward and Mapes (1971)
Ephedra foliata	Female gametophyte	MS	2,4-D 4.5 + CM 10%	Konar and Singh (1979)
Euphorbia tirucalli	Stem	MS	2,4-D 4.5 + NAA 10.7	Misawa *et. al.* (1982)
Ginko biloba	Pollen	S	2,4-D 2.71-27.12 + CM 10-40%	Lamport (1964)

Species	Explant	Medium	Growth regulators (μM)	Reference
Glycine max	Cotyledon	MS	2,4-D 22.6 - 181	Amberger *et. al.* (1992)
Gossypium hirsutum	Hypocotyl	S	NAA 10.7 + BAP 0.4 + AdS 217.2	Smith and Price (1978)
Hawarthia planifolia	Leaf	LS	2,4-D 0.9 + K 4.7	Wessels *et. al.* (1976)
Hordeum vulgare	Root	B5	2,4-D 9.4	Gamborg and Eveleigh (1968)
Helianthus sp. L.	Hypocotyl	MS	BAP 2.22 + NAA 0.27	Punia and Bohorova (1992)
Lemna gibba.	Frond	S	2.4-D 45.3 - 2IP 4.9	Chang and Chiu (1978)
Lithops leslief	Leaf	MS	2,4 D 4.5 + 46.5	Rogers and Linebeyer (1981)
Mammillaria elongata	Tubercle	S	2,4-D 27.1 + K 4.7 9.3 (or) 2iP 4.9-9.8	Johnson and Emino (1979)
Mangifera italica	Nucellus	S	2,4-D 9	Litz *et. al.* (1984)
Medicago sativa	Leaf, hypocotyl, or cotyledon	S	2.4-D 4.5 - K 0.93 + Ad 7.4	Atanassov and Brown (1984)
Nicotiana tabacum	Leaf	MS (or) B5	2,4-D 4.5	Gamborg *et. al.* (1979)'
Oryza sativa	Immature embryo	S	2,4-D 9	Grimes and Hodges (1990)
Pennisetum americanum	Inflorescence	S	2,4-D 11.3	Vasil and Vasil (1982)
Petunia hybrida	Leaf	S	2,4-D9 + K 1.2	Shneyour *et. al.* (1984)
Phoenix dactylifera	Shoot-tips or embryo	S	2,4-D 440 + 2iP 14.5	Tisserat (1984)

Species	Explant	Medium	Growth regulators (μM)	Reference
Pinus taeda	Stem	LM	2,4-D 22.6	Litvay *et. al.* (1985)
Pisum sativum	Root	S	2, 4-D 6.8 + NAA 0.5 + IAA 5.7 + K 1.2	Veliky and Martin (1970)
Pteris vittata	Rhizome	S	2,4 D9 + CM 10%	Kshirsagar and Mehta (1978)
Saccharum sp.	Leaf	MS	2,4 D 2.3 - 13.6	Ho and Vasil (1983)
Sorghum bicolor	Leaf	MS	2,4-D9 + K 0.5	Wernicke and Brettel (1980)
Theobroma cacao	Cotyledon	B5	2,4-D 4.5 + K 0.5	Tsai and Kinsetter (1981)
Triticum aestivum	Immaturembryo	S	2,4-D 9	Redway *et. al.* (1991)
Vitis vinifera	Stem, leaf, or fruit	S	NAA 0.5 + K 0.9 + CM 15%	Lie and Yic (1989)
Yucca schidigera	Hypocotyl	MS	2,4-D 4.5	McCarthy and Staba (1985)
Zamia pumila	Embryo	B5	2,4-D 2.3+ K 4.7- 18.6	Chavez *et. al.* (1992)
Zea mays	Immature embryo	S	2.4-D 4.5 - 6.8	Armstrong and Phillips (1988)

MS = Murashige and Skoog medium; LS - Linsmaier and Skoog medium; B5 = Gamborg's B5 medium; Wh = Whit's medium; S = Specific medium formulation or modified for the study reported : 2,4-D = 2.4 - Dicholorophenoxyacetic acid : NAA = 1 naphthaleneacetic acid; IAA = indole - 3 acetic acid; BAP = 6-benzylaminopurine; K = Kinetin; Ad = adenine; AdS = adenine; 2iP = N-isopenteny laminopurine; CM = coconut milk; Pic = picloram.

MEASUREMENT OF GROWTH PARAMETERS

Growth of callus and cell suspension cultures can be monitored by increase in fresh or dry weight or increase in cell number. For cell suspension cultures, increase in packed cell volume (PCV) is also a good indicator of growth.

FRESH AND DRY WEIGHT MEASUREMENTS

For callus culture, transfer the entire callus (scrape off the medium) to a pre weighed weighing boat (or container) and determine the fresh weight. For cell suspension cultures, collect the cells on a pre-weighed nylon membrane and determine the fresh weight (determine the weight of water retained by the membrane separately and subtract this amount from the measured fresh weight). Alternatively, transfer the entire contents of the cell culture flask to a pre-weighed centrifuge tube. Spin the tube for 5-10 minutes at 200 g. Examine the supernatant for cells.. If cells are present in the supernatant, repeat centrifugation until the supernatant is free from cells. Then carefully pipette out the entire medium without disturbing the pellet. Weigh the centrifuge tube with cells to determine the fresh weight. After measuring the fresh weight, dry the samples in an oven at 60°C Until no change in dry weight is observed.

Increased in fresh weight can also be measured without sacrificing the samples at the beginning of an experiment. Transfer callus or cells to pre weighed Petri dishes or culture flask containing the medium. Weight the Petri dish or the flask again and determine the weight of callus or cells added. At the end of an experiment, remove the entire callus or cells from a suspension culture and determine the final weight.

INCREASE IN CELL NUMBER

A haemocytometer can be used for determining the cell numbers in a fine suspension culture. Callus cultures and suspension cultures consisting of cell aggregates have to be macerated prior to counting the cell numbers. An ideal macerating fluid for callus or cell aggregates from suspension cultures consists of equal volumes of 10% chromic acid and 10% nitric acid. Treat cell clumps for 5-30 minutes in this mixture at 60°C. After cooling, shake the container vigorously to loosen the cell clumps. Alternatively, add 1% (w/v) pectinase or macerozyme to the culture medium and incubate flasks overnight on a shaker in the growth room under normal culture conditions.

PACKED CELL VOLUME

Transfer the entire contents of the flask to a graduated centrifuge tube. Spin the tube at 200 g for 5-10 minutes or until supernatant is free from cells. Determine the volume of the pellet after centrifugation. Pakced cell volume is expressed as a percentage of the volume of the pallet to the entire culture volume.

CELL VIABILITY

It is necessary to determine periodically the viability of cells in culture, especially if you face problems with respect to growth of the culture. Microscopic observation of cytoplasmic streaming in cells is a good indicator of cell viability, however, cell viability can also be easily and reliably evaluated using vital stains like fluorescein diacetate (FDA) stain and Triphenyl tetrazolium chloride (TTC) assay whose procedures are given below :

FLUORESCEIN DIACETATE VITAL STAIN

The fluorescein diacetate (FDA) vital stain is dependent upon the ability of esterases in viable cells to cleave the stain, which then fluoresces yellow/green under a UV microscope.

Procedure

1. Make up a stock solution of 0.1% (w/v) FDA in acetone. This may be stored at 4°C. Add a few drops of stock to 10 ml of liquid culture medium.
2. Add a few drops of FDA stain to cells on a microscope slide. Allow the stain to penetrate for a few minutes, place a cover-slip over the cells, and examine under UV fluorescence microscope using a blue/violet filter.
3. Count the number of fluorescing cells in a dark field of view and count the total number of cells in the same field under bright illumination.
4. Repeat for at least five fields of view and for each calculate :

$$\%\ \text{viability} = \frac{\text{Number of fluorescent cells}}{\text{Total Number of cells}} \times 100$$

THE TRIPHENYL TETRAZOLIUM CHLORIDE (TTC) ASSAY

This is a useful alternative for larger specimens Dehydrogenase activity of viable cells reduces TTC to a red formazan product which is measured spectrophotometrically.

PROCEDURE FOR TTC TEST

Reagents

- 0.6% (w/v) TTC in 0.05 M Na_2HPO_4/KH_2PO_4
- 95% (v/v) ethanol buffer pH 7.4 + 0.05% (v/v) Tween 80

METHOD :

1. Place 50-100 mg (fresh weight) plant tissue in a test tube, add 3 ml of TTC reagent and infiltrate under vacuum for 10-20 min.
2. Incubate at 28°C overnight.
3. Add 7 ml 95% ethanol and extract the coloured complex by boiling in a water-bath for 4 min.
4. Mix the samples throughly and centrifuge at 500-1000 g to precipitate the cell debris.
5. Measure the absorbance of the supernatant at 490 nm.
6. Express viability as absorbance$^{490\ nm}$/g fresh weight.

REFERENCES

1. Amberger, L.A. Palmer, R.G. and Sh Shoemaker, R.C. (1992). Crop. Sci. 32, 1103.
2. Armstrong, C.L. and Phillips, R.L. (1988). Crop Sci. 28, 363.
3. Atanassov, A. and Brown, D.C.W. (1984). Plant Cell Tiss. Organ Cult., 3, 149.
4. Aziz, H.A. and McCown, B.H. (1985). HortScience, 20, 540.
5. Chavez, V.M, LItz, R.E., and Norstog, K. (1992). Plant Cell Tiss. Organ Cult., 30, 90.
6. Dodds, J.H. and Roberts, Z.W, 1985 Experiments in Plant Tissue Culture (Second Ed.). Cambridge Univ. Press, Cambridge.
7. Dunstan, D.I. and Short, K.C. (1977). Physiol. Plant., 41, 70.
8. Gamborg, O.L., Shyluk, J.P., Fowke, L.C., Wetter, L.R., and Evans, D. (1979). Z. Pflanzenphysiol., 95, 255.
9. Gamborg, O.L. and Eveleigh, D.E. (1968). Can. J.Biochem. 46, 417.
10. Ghang, W. and Chiu, P. (1978) Z. Pflanzenphysiol. 89, 91.
11. Grimes, H.D. and Hodges, T.K. (1990). J.Plant Physiol., 136, 362.
12. Hankes, P.A.A., Krijbolder, L., Libbenga, K.R., Wijnsma, R., Aremge, T.N., and Verpoorte, R. (1985). Plant Cell Tiss, Organ Cult. 4, 199.
13. Havel, L. and Kolar, Z., (1983). Plant Cell Tiss. Organ Cult., 2, 349.
14. Ho, W.J. and Vasil, I.K,. (1983) Ann. Bot., 51, 719.
15. Hooker, M.P. and Nabors, M.W. (1977). Z. Pflanzenphysiol., 84, 237.

16. Johnson, J.L. and Emino, E.R. (1979). Hortiç Sci., 14, 605.
17. Konar, R.N. and Singh. M.N. (1979) Z.Pflanzenphysiol., 95, 87.
18. Kshirasagar, M.K. and Mehta, A.R. (1978). Phytomorph., 28, 50.
19. Lai, P.C. and Yie, S.T. (1980). J.Agric. Res. China, 29, 157.
20. Lamport, D.T.A. (1964). Exp. Cell Res., 33, 195.
21. Lindemann, E.G.P., Gunckel, J.E., and Davidson, O.W. (1970). Am. Orchid Soc. 'Bull., 39, 1002.
22. Litvay, J.D., Verma, D.C. and Johnson, M.A. 1985. Plant Cell Rep, 4 : 325.
23. Litz, R.E., Knight, R.J., and Garzit, S. (1984). Sci. Hortic., 22, 233.
24. MacCarthy, J.J.and Staba, E.J. (1985). Ann. Bot. 56, 205.
25. Misawa, N., Yamano, Y., Ohyama, K., and Komano, T. (1982). In Proc. 5th Int. Cong. Plant Tiss. Cell Cult., Jap. Assoc. Tiss. Cult., Tokyo. Pp. 607-608
26. Negrutiu, I., Beeftink, F., and Jacobs, M. (1975). Plant Sci. Lett. 5, 293.
27. Ozias-Akins, P., Anderson, W.F., and Holbrook, C.C. (1992). Plant Sci., 83, 103.
28. Paterson, K.E. and Rost, T.L. (1979). Am. J.Bot., 66, 463.
29. Phillips, R. and Dodds, J.H. 1977. Planta, 135 : 207-212.
30. Punia, M.S. and Bohorova N.E. 1992. Plant Sci, 87 (1) : 79-83.
31. Redway, F.A., Vasil, V., and Vasil, I.K. (1990). Plant Cell Rep., 8, 714.
32. Rogers, S.M., D. and Lineberger, R.D. (1981). Hort Science, 16, 426.
33. Shekhawat, N.S., Gordon, P.N., and Galston, A.W. (1984). *In Vitro* Cell. Dev. Biol., 20, 707.
34. Shneyour, Y., Zelcer, A., Izhar, S., and Beckmann, J.S. (1984). Plant Sci. Lett., 33, 293.
35. Smith, R.H. and Price, H.J. (1978) Plant Physiol., 61, 19.
36. Sondahl, M.R., Nakamura, RT., Medina-Filho, H.P., Carvalho, A., Fazuoli, L.C. and Costa, W.M. (1984). In Handbook of plant cell culture (ed. P.V. Ammirato, D.A. Evans, W.R. Sharp, and Y. Yamada), Macmillan Publishing Co., New York. Vol. 3, p. 564-590.
37. Stafford, A., Smith, L., and Fowler, M.W. (1985). Plant Cell Tiss. Organ Cult., 4, 83.
38. Steward, F.C. and Mapes, M.O. (1971). Bot. Gaz, 132, 65.
39. Thorpe, T.A. 1980. In : Perspective in Plant Cell and Tissue Culture. Internate Rev. Cytol., Suppl. 11A (I.K. Vasil ed.) Acad. Press, New York., PP 71-111
40. Tisserat, B. (1984). In Handbook of plant cell culture (ed. W.R. Sharp, D.A. Evans, P.V. Ammirato, and Y. Yamada), Macmillan Publishing Co. New York. Vol. 2, pp. 505-545.
41. Tsai, C.H. and Kinsellar, J.E. (1981). Ann. Bot., 48, 549.
42. Vasil, V. and Vasil, I.K. (1982). Am. J. Bot., 69 1441.

43. Veliky, I.A. and Martin, S.M. (1970). Can. J. Microbiol., 16,223.
44. Wernicke, W. and Brettel, R. (1980). Nature, 287, 138.
45. Wessels, D.C.J., Groenewald, E.G., and Koeleman, A. (1976). Z. Pflanzenphysiol., 78, 141.
46. Wilmar, C. and Hellendoorn, M. (1968). Nature, 217, 369.

6

OVULE CULTURE IN WHEAT

In case of difficult interspecific or intergeneric crosses, that have low success rate, ovules after fertilization have been cultured successfully to obtain mature embryos/seed. Depending upon, When the embryo aborts, the ovules have to be excised any time soon after fertilization to almost developed fruits, which sometimes may be lost due to premature abscission. However, the ovule culture is mainly tried only in those cases, where embryo abort very early, and embryo culture is not possible due to difficulty of its excision at a very early stage. The detailed procedure for wheat ovule culture is as below :

Protocols

1. Emasculate the wheat spikes 1-3 days before anthesis, using only the spikelets of synchronously developed middle part of the spike and the two outer florets from end spikelet.
2. When stigma become receptive, pollinate each floret using 2-3 freely dehisced anther for each floret.
3. Between 6 and 24 h after pollination, harvest the inflorescances/ spikes & the ovaries are surface sterilized for 5 min in a 5% (w/v) Tween 20.
4. Give four to six washes of sterilized distilled H_2O.
5. Ovules are excised using fine-tipped forceps, a scalpel and preparation microscope. Cut the pericarp along one side from the chalazal to the micropylar end and across the stigma. Lift the pericarp carefully from the abaxial side of the ovule and subsiquently isolate the ovule.
6. Culture the ovules with the adaxial side downwards on modified OC medium (Kumlehn and Nitzsche, 1995) with 85.5 g/l sucrose,

80 mg/l NH_4SO_4, 0.5 mg/l Kinetin, and, unless stated otherwise solidified with phytagel, filter sterilized at pH 5.7.

7. Incubate the cultures at 24°C in the dark for 3 weeks for callus development.
8. Transfer well developed calli to the medium consisted of B5 salts and vitamins (Gamborg *et. al.* 1968), organic acids of Kao and Michayluk (1975), 250 mg/l casein hydrolysate, 34.2 g/l sucrose, 250 mg/l xylose, 7 g/l phytagel, 6.5 mg/l IBA and 0.1 mg/l thidiazuron pH 5.7, autoclave sterilized.
9. Subculture at an interval of 25 days till development of embryos and subsequent plant regeneration/germination of embryos, This method would be used in an embryo rescue technique in plant breeding.

REFERENCES

1. Gamborg, O.L., Miller, R.A. and Ogima, K. 1968. Exp. Cell Res. 50 : 151 - 158.
2. Kao, K.N. and Michayluk M.R. (1975). Planta, 126 : 105 - 110.
3. Kumlehn, J. and Nitzsche, W. 1995. Plant Sci. 1111 : 107 - 116.
4. Kumlehn, J., Schieder, O., and Lorz, H. 1997. Plant Cell Reports, 16 : 663 - 667.

7

ANTHER CULTURE FOR HAPLOID AND DIHAPLOIDS (DH) PRODUCTION IN SUNFLOWER (*HELIANTHUS ANNUUS* L)

The term "Haploids" refers to those plants which possess the gametophytic number of chromosomes in their sporophytes. Haploids and double haploids/dihaploids (DH) could be of great value for breeding and genetic analysis of crop plants. As breeding of crop plants requires large population for selection, the anther culture system, in which main spore from an anther forms a callus from which haploid plants can be regenerated, is potentially the ideal system for generating dihapoloids plants that are needed. The main advantage of the *in vitro* production of haploids is to produce homozygous plants (inbred lines) and isogenic lines within a year as compared to long inbreeding methods which might take 4 to 6 years.

Protocols

1. Collect anthers of uniuncleate microspore stage from field grown plants of sunflower.
2. Anthers should be washed throughly in soup solution and rinsed with distilled water.
3. After this, anthers are surface sterilized in 0.1% mercuric chloride solution ($HgCl_2$) by adding 2-3 drops of Tween 20 for 3-4 minutes followed by repeated washing 6-7 times in sterilized distilled water in a laminar air flow.
4. Keep the sterilized anthers is sterilized petri dishes for their subsequent plating.
5. Sterilized anthers are plated in petri dishes containing AnS3 medium (MS medium supplemented with 0.5 mg/l NAA, 0.5 mg/l

BAP, 1.0 mg/l each of thiamine HCl, Calcium Panthothate, mesoinositol, pyrodoxine HCl, Nicotinic acid and 0.01 mg/l biotin, 0.1 mg/l Vit B/2, 60g/l agar, PH 5.8). Sealed the petridishes with parafilm.

6. Incubate the cultures at 25°C ± 1°C under dark condition initially for 15-20 days & than transfer the cultures under light condition 16/8 h. photo period.
7. Subculture in the same medium at an interval of 25-30 days for callus growth.
8. Transfer the calli into flasks containing regeneration medium MSZ MS (supplemented with 1.0 mg/l Zeatin, 30g/l sucrose, 8.0 g/l agar and pH 5.8.
9. Subculture the calli on MSZ medium at an interval of 30 days till the haploid plantlets regenerate.
10. Separate the haploid plants from each other and transfer them to MS basal medium for rooting.
11. Treat the shoot tip of the haploid plants with calchicine @ 0.05% for the induction of diploid shoots for their further growth & flowering.
12. Harvest the mature seeds for raising the next generation of DH lines.

REFERENCES

1. Punia, M.S. 1997. Proc. Intl. Conf. Sustainable Agriculture for Food, Energy and Industry, Braunschweig (Germany) June, 1997 In Press (Abstract pp. 123).

8

EMBRYO CULTURE FOR CROP IMPROVEMENT

Henning (1904) first cultured embryos of two cruciferous genera that opened up the field of *in vitro* embryo culture with significant application in crop improvement programmes. Since then, embryo culture has been used to explore the nutritional and physical requirement for embryonic development, to bypass seed dormancy which may shorten the breeding cycle, to test seed viability, to provide microcloning source material and to rescue immature hybrid embryo from incompatible crosses.

A procedure for wheat (*Triticum aestivum*) embryo culture is described below:

PROCEDURE

1. Collect the midsections of spikes, containing 15-20 developing seeds, 14 days post-anthesis and disinfect by dipping in 95% ethanol for 2 min and 3-5 min in 0.1% $HgCl_2$ followed by three to five washes with sterile distilled water.
2. Dissect out the immature embryos (1-1.5 mm in size) from developing seed in aseptic laminar hood under a dissecting microscope on an ethanol swabbed surface, or even without microscope.
3. For dissection, peel back the surrounding leaf-like structures (glume, lemma, palea) from the developing grain and remove the smooth green grain.
4. Holding the grain, smooth side up, with aseptic forceps, and with the coleorhizal end towards the viewer, make a transverse incision with an aseptic scalpel across the grain at midpoint. Make a second incision along the side of the grain towards the coleorhizal

end. Peel back the epidermal flab with the forceps and carefully lift out the white ovalshaped embryo with the tip of the scalpel.

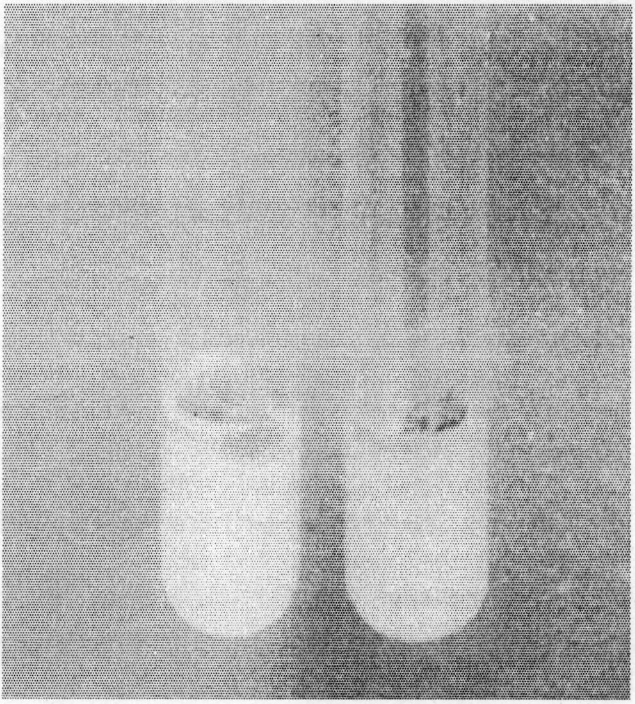

Fig. 8.1 Callus from Wheat Embryo

5. Plate/inoculate the embryos scutellum down on callus initiation MS medium supplimented with 2.0-5.0 mg/l 2,4-D (4-6 embryos in a petri dish of 60 × 15 mm size). Sealed the petri dished with parafilm and incubate them in growth room at a temperature of 25° ± 2°C under 16/8 hours photoperiod for callus development (Fig 8.1).
6. Subculture the calli at an interval of 30 days.
7. Transfer well developed calli on MS medium supplimented with 0.5 mg/L IAA and 0.1 mg/L 6-BAP in 250 ml flasks/culture vessels for differentiatio/embryogenesis and plantlet regeneration (Fennel *et. al.* 1996).
8. Separate the regenerated plantlets and place them on MS basal medium individually in test tubes for their rooting.
9. Remove the fully rooted plants from test tubes, wash the roots gently with water to remove the traces of medium/agar and transfer to the pots containing soilrite mixture under high humidity conditions to avoid descication.

10. Give water to the plants whenever required & transfer them to green house after two weeks for their further growth and flowering.

Note: Embryo culture protocol may slightly vary from species to species. Depending upon the species in questions, the protocols have been modified. Therefore, it is suggested to follow the suitable protocols developed for various species.

EMBRYO RESCUE FROM INCOMPATIBLE HYBRID CROSSES

In breeding, interspecific and intergeneric crosses are frequently carried out to transfer valuable genes from wild species to cultivated species. Incompatibility is often encountered in such crosses and results in shriveled seeds containing aborted embryos. In some interspecific or intergeneric crosses, the failure of endosperm development eventually leads to the starvation and abortion of the embryo. The culture of hybrid embryo on an artificial nutritional media prior to abortion may circumvent these string postzyotic barriers called embryo rescue technique. Therefore, embryo culture has been utilized frequently in rescuing novel hybrid plants which otherwise are difficult to obtain.

The stage of development of the embryo at the time of culture is important. Embryo culture is most frequently used in rescuing incompatible crosses in breeding cereal crops. Gill et al. (1981) produced viable *Aegilops squarrosa* × *Triticum boeticum* hybrid plants by this technique, culturing 5-10 days old developing embryos after pollination. The procedure and stage of isolating/dissecting developing embryo for its rescuing and culture on nutritional media varies greatly with species-used in the hybridization programme.

Therefore, optimum time of dissecting/removing embryo from an incompatible cross to rescue it followed by suitable culture protocols are important for its success. Hence, efforts should be made to identify the proper conditions and requirements for culturing embryos of progressively younger stages in the major crop plants.

REFERENCES

1. Fennell, S., Bohorova, N., Ginkel, M.V.; Cross, J., and Hoisington, D. 1996. Theor. Appl. Genet, 92 : 163-169.
2. Gill, B.S., Waines, J.G. and Sharma, H.C. 1981. Plant Sci. Lett., 23 : 181-187
3. Hanning, E. 1904. Bot. Ztg. 62 : 45-80.

9

SOMATIC EMBRYOGENESIS IN ALFALFA (*MEDICAGO SATIVA*)

Somatic embryogenesis is the process by which the somatic cell gives rise to bipolar structures which develop into whole plants without gametic fusion. This process strongly depends on the plant genotype, the type and physiological state of the explant, the composition of the culture media and the culture conditions. Somatic embryogenesis could be "direct" when the embryogenic structures develop directly from the initial explant, or "indirect" through a callus stage.

Briefly the advantages of direct somatic embryogenesis could be summarised as:

1. It omits the callus stages and allows the shortening of the process of somatic embryogenesis.
2. It permits the avoidance of secondary changes occurring during the process of dedifferentiation.
3. It provides ground for further morphological, biochemical and molecular studies of the nature of somatic embryogenesis.

Plant Material

Use *in vitro* raised 30 days old sterile shoot culture of alfalfa established from seeds. In case of seed as starting material, the following sterilisation procedure should be used.

(a) *Pre-sterilisation :*

Wash the seeds in detergent solution for 15 min.

Submerge the seeds in 70% alcohol solution for 1 min.

(b) *Sterilisation:*

submerge the seeds in 0.2% $HgCl_2$ for 10 min.

(c) *Post sterilisation :*

wash 6-7 times in sterile distilled water.

Use sterile forceps and transferring the seeds into tubes with culture medium free of growth regulators (MSo). Propagate the seedlings via cuttings on the same medium at 25 ± 1°C under a 16/8 hours photoperiod.

MEDIA

For all plant culture media, macro-elements are prepared as 100 fold concentrated stock, micro-elements and vitamins are 1000 fold concentrated. The pH of the media is adjusted to 5, 8 with NaOH. The composition of the used media are given as follow :

MSo = MS macro-elements, MS microelements, MS vitamins, 3% sucrose and 0.7% agar. (Murashige and Skoog, 1962).

B_5o = B_5 macroelements, B_5 microelements, B_5 vitamins 2% sucrose. (Gambourg, *et.al.*, 1968)

Blaydes = BI macroelements, B_l microelements, B_l viatmins, 3% sucrose and 0.7% agar. (Blaydes, 1966).

B_5II is modified B_5 medium supplemented with 500 mg/I inositol, 500mg/I casein hydrolisate, 2g/l 2,4D, 0,2 mg/1 kinetine, 1mg/1 adenine, 10 mg/1 glutamine and 0.7% agar.

B_5IV is B_5II supplemented with 4 mg/l 2,4D + 0.2 mg/l KIN

B_53M = B_5o supplemented with 2.5% PEG 6000, 3% maltose instead of sucrose.

Protocols

A. Indirect somatic embryogenesis

I. Induction of callus and embryogenic suspension culture

1. Take young but fully expanded leaves of a shoot culture under sterile conditions and put them in a Petri dish. Cut the leaves in a small pieces (4-5mm) with a sharp scalpel and put the explants onto B_5II solid medium. The wounded side should contact the agar. Incubate dishes for about 35-40 days

under the standard conditions (25 ± 1°C and 16/8 hours photoperiod).

2. For initiation of embryogenic suspension culture, transfer induced friable rapidly growing callus into agitated liquid medium of the same composition as that used for callus growth. Agitation rate should be in the range of 80-100 r.p.m. Use approximately 500mg of the callus and inoculate it into a 100ml flask, containing 25ml medium.

3. For embryo conversion to plantlets, plate the induced somatic embryos in cotyledon stage on solid B_5o or Blaydes medium.

Note: Cell suspension cultures usually require regular subculture at more frequent intervals than the callus cultures. Subculture involves the aseptic transfer on a suitable volume of inoculum to fresh medium.

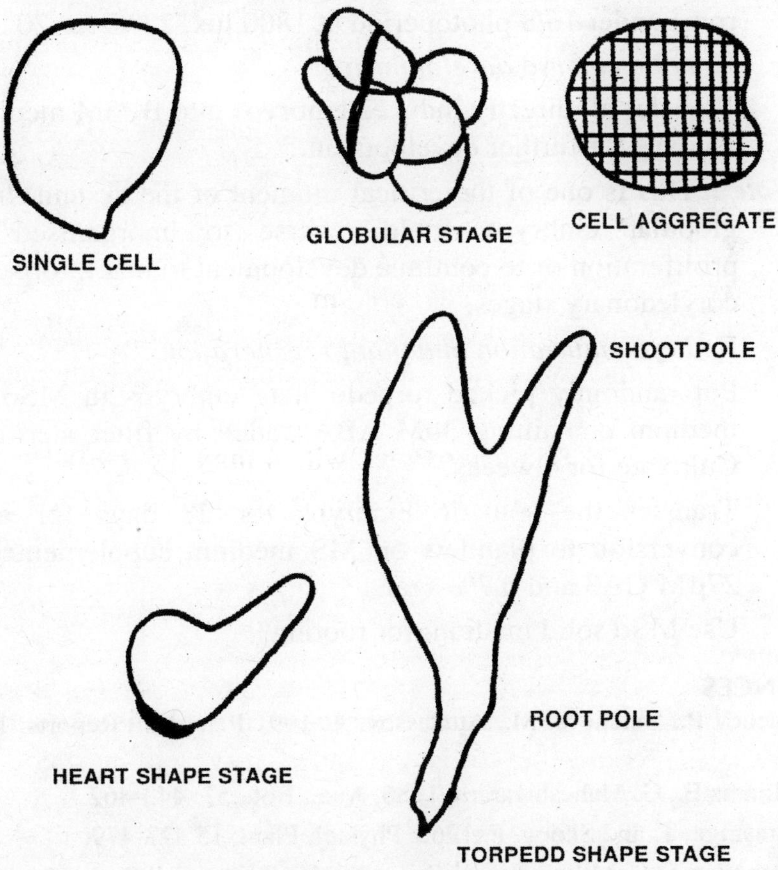

Fig. 9.1 Stages of embryogenesis

B. Direct somatic embryogenesis

I. Induction of direct somatic embryogenesis

1. *Explant Preparation*

   The most important step in developing the system for direct somatic embryogenesis is to choose the suitable explant. Use young trifoliate leaves separated from 30 days old cuttings. Cut the explant into small pieces in a liquid B_5o medium. Wash several times the cut explants with the same fresh medium.

2. *Induction of direct somatic embryos :*

   After the last washing inoculate the explants into 100 ml flasks, containing a definite volume of B_4IV induction medium. Add 100mg of the leaf material to 100 ml of the medium. Maintain the cultures on horizontal shaker (100 rpm) under 16/8 photoperiod at 1800 lux, 22°C for 20 days.

3. *Somatic embryo development*

   Transfer all directly induced embryos into B_5/3M medium for 25 days for further development.

   Note : This is one of the critical moment of the system. Induced globular embryos could reverse to unorganised callus proliferation or to continue development to heart, torpedo and cotyledonary stages.

4. E*mbryo maturation and plant regeneration :*

   Put randomly picked torpedo state embryos in MSo liquid medium containing 30M ABA added by filter sterilisation. Cultivate for 4 weeks.

5. Transfer the somatic embryos for 28 days for embryo conversion to plantlets on MS medium supplemented with 27μM GA3 and 0.7% agar.

6. Use MSo solid medium for rooting.

REFERENCES

1. Denchev P., Velcheva, M., Atanassov, A. 1991 Plant Cell Reports, 10, 338-341.
2. Williams E., G. Maheshwaram, 1986, Ann,. Bot., 57: 443-462.
3. Murashige, T. and Skoog, F. 1962. Physiol. Plant, 15:473-479.
4. Gambourg, O.L, Miller, R. and Ojima, K. 1968 Exp. Cell Res ; 50 : 148.
5. Blaydes, D.F. 1966. Physiol., Plant, 19 : 748-753.

10

Production of Synthetic Seeds

Synthetic or artificial seeds are the living propagating material (seed like structure) deried from *in vitro* produced somatic embryoids after encapsulation by a hydrogel. The preserved embryoids are termed as synthetic seeds. Somatic embryoids are indentical with zygotic embryos that give rise to plants only under controlled laboratory conditions. Somatic embryos are without seed coats, develops from callus tissue through somatic embryogenesis *in vitro*. Such seeds are contaminated with microbes and desicate quickly when they are subjected to field conditions. Therefore, to overcome such problems, they are encapsulated by a protective gel like calcium alginate. Encapsulated embryoids can resist unfavourable field conditions without dessication. The seeds so developed behave like a true seeds and are used as a substitute of natural seeds (Fig. 10.1).

The following steps are taken to produce synthetic seeds :

1. Establishment of callus culture.
2. Induction of Somatic embryogenesis.
3. Maturation of somatic embryos.
4. Encapsulation of somatic embryos.
5. Evaluation of embryoids and plant conservation.
6. Planting in field/green house.

Protocol

Production of encapsulated dried somatic alfalfa embryos

1. Plunge whole petioles which have had two of the three leaflets removed into a saturated, pre-filtered solution of calcium hypochlorite for 10 min. Agitate gently and periodically during the treatment. Wash three times with sterile distilled water.

2. Aseptically trim 1-2 mm from the end of each petiole, cut into approximate 1 cm sections, and float on sterile distilled water.
3. Transfer trimed 1 cm petioles sections onto callus induction B5h, medium (Table 1). To avoid desiccation, float the petioles in a small amount of sterile distilled water in a Petridish during cutting and while awaiting transfer.
4. Plate five to ten petioles on 20 ml solidified B5h medium contained in a 20 × 600 mm Petridish wrapped with Parafilm prior to culture at 25°C under a 12-16 h photoperiod of fluorescent light at 50 E $m^{-2}.s^{-1}$. Culture for 21-28 days until early globular staged embryo's are visible on the surface of the callus.
5. Aseptically transfer five to seven well developed calli (20-25g) to 45 ml liquid B5m medium (Table 1) contained in a 125 ml Erlenmeyer flask. Plug flasks with autoclavable foam plugs, cap with aluminium foil, and culture on an orbital shaker at 130 r.p.m. under the above conditions. Cell dispersion and growth should be evident within three days.
6. Select flasks with rapidly growing cell suspensions, 8-12 days old with a clear 'growth ring' of cells and young green globular embryos on the mid-upper portion of the flask.
7. Pass the cell suspension sequentially through a 1000μm and 200μm nylon screen, and retain the cells which accumulate on the 200μm screen.
8. Transfer about 1 g of selected cell suspension to an aseptic 50 mm diameter 200μm mesh nylon screen on 20 ml of solidified, BOi2Y medium (Table 1) and spread evenly over the surface of the nylon screens.
9. In order to reduce secondary embryo formation and induce tolerance to desiccation, culture embryos on BOi2Y medium supplemented with 10^{-5} M absicisic acid by transferring the nylon screen containing the cell suspension and developing embryos onto the BOi2Y-ABA medium at day 11. The glutamine and sucrose incorporated in this medium are beneficial to embryo maturation.
10. After 7-14 days on ABA-containing medium and after embryos have acquired a cream-white colour, harvest the embryos by collecting on a 200μm nylon filter and washing off undeveloped cells and residual medium with sterile distilled water.

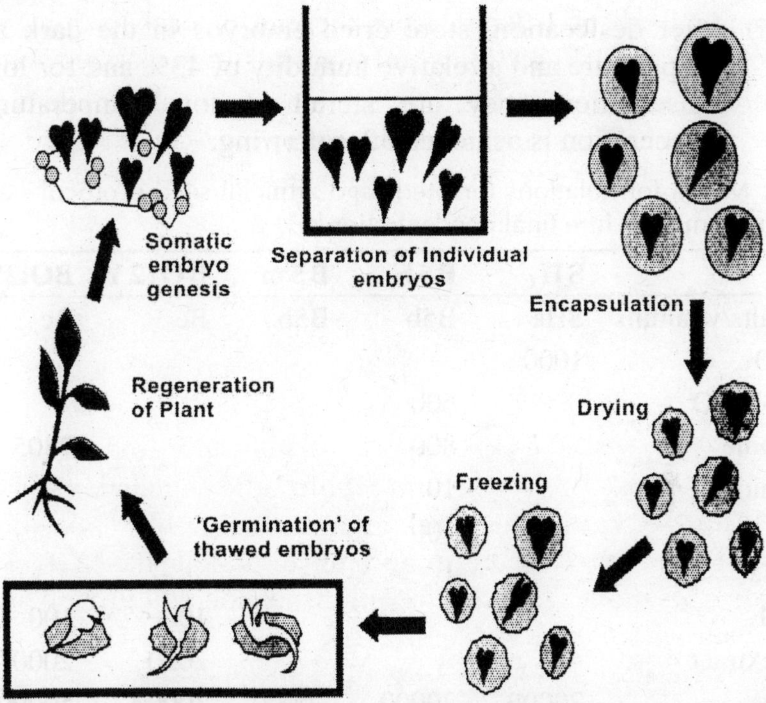

Fig. 10.1 Production of Synthetic Seeds

11. To Protect the embryos or synthetic seeds, encapsulation is done with various types of hydrogels and their complexing agents. Most commonly used gels are :-
    (i) Calcium alginate (0.5-5.0% W/V) with its complexing agent calcium nitrate (30-100μM);
    (ii) Sodium alginate (2.0% W/V) with gelatin (5.0% W/V) and use calcium chloride (30-100μM) as complexing agent, and
    (iii) Locust Beam Gum (0.4-1.0% W/V) with ammonium chloride (500μM) as complexing agent. After encapsulation, the artificial or synthetic seeds are directly used for raising the crop in the field.
12. *For conservation or for long/short term storage, the following steps can be used:*
    (i) Transfer, without wrapping, Petri dishes containing selected embryos spread on damp filter paper to desiccator with a controlled relative humidity of 97%.
    (ii) Every 24-48 h, transfer the dishes to desiccators with progressively lower relative humidity levels, as outlined in Table 2.

(iii) After desiccation, store dried embryos in the dark at room temperature and a relative humidity of 43% and for long term conservation, they are stored at low temperature. The regeneration is achieved after thawing.

Table 1. Media formulations for Medicago artificial seed protocol - additions are given as mg per litre final concentration.

	SH_f	B5h	B5m	BOi2Y	BOi2Y-ABA
Base salts/vitamins	SHa	B5b	B5b	Bc	Bc
NH_4NO_3	1000				
$CaCl_2$-$2H_2O$		600			
Glutamine		800			7305
Glutathione		10			
Serine		100			
Adenine		1			
Inositol				100	100
Yeast extract				2000	2000
Sucrose	20000	30000	20000	30000	50000
2,4-D		1.0	1.0		
Kinetin		0.2			
Naphthalenacetic acid			0.1		
Indole-3-acetic acid	0.02				
Isopentenyladenine	0.02				
Abscisic acid					13.2
Bacto agar (Difco)	10.0	9.0		10.0	10.0
pH	5.9	5.5	5.5	5.9	5.9

a. Schenk and Hildebrandt medium (1972), see Appendix 1.
b. Gamborg et. al. B5 medium (1968), see Appendix 1.
c. Blaydes medium (1966).

Table 2 : Relative humidity conditions used during desiccation.

Desiccator	Saturated compound	Theoretical % RH (8-11)	Measured % RH
1.	K_2SO_4	97.5	92
2.	$Na_2CO_3.10H_2O$	87	92

Desiccator	Saturated compound	Theoretical % RH (8-11)	Measured % RH
3.	NaCl	75.5	88
4.	NH_4NO_3	62.5	79
5.	$Ca\ (NO_3)_2.4H_2O$	50.5	52
6.	$K_2CO_3.2H_2O$	43	48

REFERENCES

1. Redenbaugh, K. (ed.) (1992). Syn. Seeds; Application of synthetic Seeds to Crop Improvement, CRC Press, Boca Raton, Florida.
2. Brown, D.C.W. 1995. In: Plant Cell Culture - A Practical Approach (Dixon, R.A. and Gonzales, R.A. Eds). Oxford University Press, Oxford. New York, Tokyo, p. 138-145.
3. Murashige, T. and Skoog, F. (1962). Physiol. Plant, 15, 473.
4. Blaydes, D.F. (1966). Physiol Plant., 19, 748.
5. Schenk, R.U. and Hildebrandt, A.C. (1972) Can. J.Bot., 50, 199.
6. Gamborg, O.L. Miller, R.A., and Ojima, K. (1968). Exp. Cell Res., 50,151.
7. Huang, B. and Keller, W.A. (1989). J.Tiss. Cult. Meth. 12, 171.
8. Brown D.C.W. Watson, E.M. and Pechan, P.M. (1993). *In Vitro* Cell. Dev. Biol. Plant, 29, 113.
9. Pechan, P.M., Bartels, D., Brown, D.C.W. and Schell, J. (1991). Planta, 184. 161.

11

In Vitro Selection for Salt Tolerance

Salt tolerance in crop plants is of an immense importance characteristic because of salinization of irrigated lands. Since traditional plant breeding methods have been slow to yield substantial improvements in salt tolerance, the alternative approach of utilizing plant cell culture and regeneration of plants from potential cell mutants has received increased attention. An added advantage in generation of salt tolerant plants through tissue culture is that the regenerated plants are clones of the original starting material that differ only in one or a few characteristics, thereby simplifying subsequent testing.

Thus, the generation of salt tolerant plants through tissue culture becomes a viable alternative to classical plant breeding in improving the salt tolerance of crop plants.

Protocols

1. Semi-Solid liquid medium is prepared and NaCl (500 mg/l) is added to the medium, pH is adjusted to 5.8.
2. Distribute the medium in 30 ml aliquots into 150 ml flasks or 50 ml aliquots into 250 ml flasks and grow the callus on NaCl containing medium.
3. Select granular friable callus for further subculture.
4. 1g of callus is inoculated to each flask.
5. Incubate the cultures on a gyratory shaker (100 rpm) for 21 days at 25 ± 2°C.
6. Pass the cell suspension through a stainless steel wire mesh (300μm) and collect filtrate in 15 ml centrifuge tubes. Examine the filterate for single cells and small aggregates (5-20 cells).
7. Centrifuge the filtrate at 100g for 5 min. and discard the

supernatant, add 5 ml of the sterilized medium to the pellet to obtain a cell suspension, and maintain its cell density at about 0.5-2.5×10^5 cells/ml. Add the cell suspension to the flasks containing with or without NaCl.

8. Incubate the cultures at 100 rpm at 25 ± 2°C and centrifuge the cultures grown in control and NaCl (500 mg/l) at 100 rpm for 5 min.
9. Transfer the pellets to fresh medium of the same composition. Incubate at 25 ± 2°C for 21 days and repeat these steps for 2-3 times to get a good growth rate of NaCl tolerant cell lines.
10. Grow cultures on 1000 mg/l NaCl and follow the same steps, as carried out at 500 mg/l NaCl till a salt concentration beyond which the cells cease to grow. It is suggested that the salt concentration should be increased gradually by repeating steps.
11. Regenerate variant cells into the surface of the regeneration medium gelled with agar by a wide mouth graduated sterile pipette or by pasteur pipette. In case of callus, inoculate 500 mg piece onto surface of regeneration (shoot/organforming) medium having 500 mg/l NaCl. Regeneration medium varies from species to species.
12. Transfer the regenerated shoots to rooting medium (MS_0).
13. After rooting, finally transfer the NaCl tolerant plantlets to soils.
14. Use NaCl solution to water the salt torerant plants. The concentration of NaCl should be the same as used in regeneration medium.

Note : This procedure can also be used to develop tolerant plants of alkali, high temperature, drought, disease and herbicide by selecting Cell lines using *in vitro* selection system.

REFERENCES

1. McCoy, T.S. 1987. Plant Cell Rep. 6;417.
2. Ibrahim, K.M., Ciollins, J.C. and Lollin, H.A 1992. Plant Cell Tissue and Organ Culture, 28-139.
3. Wincov, I. 1994 In : Plant Cell Culture - A Practical Approach (Dixon, R.A. Gonzales, R.A. eds.) Oxford University Press, Oxford, New York and Tokyo pp. 79
4. Wincov, I. and Seemann, J.R. 1990 Plant Cell Physiol. 31 : 1155.

12

In Vitro Selection for Disease/Toxin (*Verticillium* Wilt) Resistance Cell Lines in Alfalfa

Plant tissue cultures treated with toxins from disease-causing organisms are useful for examining plant-pathogen interactions and improving the disease resistance of crop plants. In particular, many of the molecular details of the hypersensitive reaction have been obtained from studies of crude or purified fungal filtrate preparation and plant tissue cultures (1). Also, pathogen preparations have been shown to contain specific or general toxins that have been used top select disease-resistant plant material (2).

In this view, selection for disease resistance is advantageous for several reasons including that:

(a) Experimental units can be maintained on defined media and in rigorously controlled environments allowing for selection of small increments in disease resistance.

(b) Cultured cells can be uniformly exposed to the selective agents, thus reducing the incidence of escapes.

(c) Culture systems maintained in small spaces can potentially replace expensive greenhouse or field testing facilities.

(d) *The disease causing agent remains confined to the laboratory. The protocols are discribed for the following stages :*

   I. Obtaining a toxic fungal culture filtrate.

   II. Measuring the effect of the toxic fungal culture filtrate on callus cultures.

   III. Regenerating plants from toxin resistant cultures.

IV. Evaluating the desease resistance of the regenerated plants.

The protocols are based on methods that were developed for studying *Verticillium* wilt in alfalfa but the general principles should be applicable to a wide variety of pathogen-plant interactions.

I. Preparation of toxin from fungal culture filtrate

In an *in vitro* selection experiment, plant tissue cultures are exposed to toxin preparations obtained from phytopathogens (3, 4, 5). These preparations can be subcellular fractions from the disease-causing organism or can be compounds released into media used to grow the pathogen.

A. Preparation of Potato Dextrose Agar (PDA)

Potato Dextrose Agar is a standard medium used for culturing bacteria, fungi, yeast and molds. The basic steps for preparing this product are listed below :

1. Suspend 39.0 g PDA powder mixture or 200 g meshed & filtered potato, 20 g dextrose and 20 g agar in 1 liter of distilled water and heat to boiling or until the medium is completely dissolved.
2. Sterilize the medium in a validated autoclave at 1 Kg/cm^2 (15 psi), 121°C. The medium should attain a temperature of 121°C for at least 15 minutes.
3. Dispense the medium as desired.

STORAGE : All medium preparations should be stored at 2-6°C. Store dry medium at room temperature (below 30°C).

B. Preparation of fungal culture filterate (6, 7)

Equipment and reagents

1. Potato dextrose agar
2. B5 media
3. Petri dishes (15 mm × 100 mm)
4. Erlenmeyer flask (250 ml)
5. Growth cabinet (temperature and light regulated)
6. Platform shaker
7. Watman 1 filter paper
8. Centrifuge capable of spinning 50 ml centrifuge tubes at 10000 g.

9. 50 ml centrifuge tubes
10. 0.22 μm filter sterilization unit.

METHOD :

1. Maintain *Verticillium alboatrum* by transferring a 5 mm square of mycelium-containing potato dextrose agar (PDA) from a 30-day-old culture to a new PDA dish every month and incubating the culture at 22°C in low light (0.5 e-m^{-2}.s^{-1}).
2. Inoculate 50 ml of liquid B5 medium (without plant hormones) in a 250 ml Erlenmeyer flask with 10-15 pieces of PDA containing mycelium cut from a two-week-old *Verticillium* culture dish.
3. Incubate the liquid cultures at 22°C in low light (0.5 e-m^{-2}.s^{-1}) with agitation (90 r.p.m.).
4. Subcultue after two weeks by adding 50ml of fresh medium to the culture and dividing it between two flasks (by six weeks the culture should consist of a uniform suspension of mycelial balls approximately 2 mm in diameter).
5. Prepare the cell-free filtrate from cultures that have been subcultured for a minimum of six weeks and a maximum of four months.
6. Collect the filtrate from cultures two weeks after their last subculture by filtering them by gravity through a Whatman No. 1 filter.

C. Plant tissue culture for *in vitro* selection

In vitro selection can be used with a population of cells that have different genetic compositions. Such as microspore or anther cultures, but most often it is used with cultures derived from somatic tissues that are, to a large extent genetically identical. In the latter case selection acts to isolate rare spontaneous mutants that show resistance to the toxic components in the culture filtrate; these mutants are usually called somaclonal variants (8.9).

Both callus the suspension cultures are useful for studies of plant pathogen interactions. Suspension cultures may be more appropriate if the objective of the experiment is to isolate resistant plant material because the smaller cell clusters that are found in these cultures are more uniformly exposed to the toxic agent.

Equipment and reagents

* Alfalfa petioles from young leaves

* Parafilm
* Laminar flow bench
* Tissue culture incubator with lights
* Sodium hypochlorite 5.6% (commercial) bleach solution 100%)
* Warning blender and sterile mini-sample blender heads.
* Sterilized distilled water
* Autoclaved 7cm Whatman No. 1 filter paper disks (may be necessary to repeat autoclaving to ensure sterility)
* Petri dishes (15 × 100 mm)
* B5h media containing B5 salts plus 1 mg/litre 2.4-D and 0.2 mg/litre kinetin (add 8% agar for solid media).

METHOD :

1. Surface sterilize the alfalfa petioles by placing them into 5.6% sodium hypochlorite in a laminar flow bench for 60 sec.
2. Rinse the petioles three times with sterilized water.
3. Remove the bleached ends and cut the petioles into 1 cm. pieces.
4. Transfer the pieces to 15 × 100 mm dishes containing B5h, and seal the dishes with Parafilm to prevent desiccation and contamination.
5. Incubate the cultures in a growth cabinet set at 25°C and a 16 h light 8 h dark photoperiod (90 e.m^{-2}.S^{-1}) for three to four weeks to allow callus to grow from the petiole pieces.
6. Add weighed amounts of callus to liquid B5h (10) in the mini-sampler blender head to give a ratio of 1 g callus/4 ml medium.
7. Homogenize the callus with three 3 sec. pulses of the Waring blender set to low speed. The homogenization allows the cells to be spread uniformly across the filter paper.
8. Add 1 ml aliquo of well-mixed callus homogenate, with a 5 ml wide mouthed pipette on to a sterile Whatman No. 1 filter placed on solid B5h medium (with or without the fungal filtrate) in 15 × 100 mm Petri dishes.
9. Spread the homogenate evenly over the top filter paper by gently swirling the dish.
10. Two hours after adding the homogenate weigh each culture without the lid.
11. Gently lift the filter paper from the Petri dish with two sterile forceps and record the weight of the dish plus the medium.

12. Determine the weight of the callus by subtracting the weight determined in step 11 from the average weight for the filter paper.
13. Seal the dishes with Parafilm and incubate under the same conditions described in step 5.
14. Repeat the weighing procedure every four days.
15. Express the callus growth in the dishes containing the culture filtrate as a percentage of the growth in the control dishes.

II. Measurement of the effect of the fungal culture filterate on alfalfa suspension culture growth and viability :

Equipment and reaents

* Three-to four week-old alfalfa callus
* Fluorescein diacetate (FDA)
* Microcentrifugal tube 1.5 ml.
* Erlenmeyer flasks
* Microcentrifuge
* B5g media
* Freezer
* 10 ml graduated centrifuge tube
* Spectrofluorometer
* Bench-top centrifuge
* Polytron homogenizer

METHOD :

1. Add approximately five petiole calli to a 125 ml Erlenmeyer flask containing 25 ml B5g media.
2. Agitage flasks at 110 r.p.m. for seven days at 25°C under 16h light 8h dark photoperiod (90 E.m^{-2}s^{-1})
3. Transfer 20 m of a seven-day-old suspension culture to a 250 ml flask containing 12 ml of B5g plus 8 ml B5 for a cycle one control culture; or to a flask containing 12 ml of B5g plus 8 ml of the fungal culture filtrate for a cycle one treatment culture with 20% (v/v) fungal culture filtrate.
4. Subculture control and treatment cultures at 12 day intervals by adding 20 ml of the previous cycle culture to 20 ml of fresh B5g media containing 20% B5 or 20% filtrate (16 ml B5g plus 4 ml B5 or 4 ml filtrate), respectively.

5. On day 0, 4, 8, and 12 of each cycle, transfer 1 ml aliquots from the filtrate and control cultures to graduated test-tubes and add 10 ml of FDA solution (5 mg/ml of acetone) to the samples.
6. Vortex the samples and allow them to stand at room temperature for 15 min; vortex again, and allow the samples to stand for another 10 min.
7. Adjust the volume of the sample to 4 ml with distilled water and centrifuge for 5 minutes at 350 g.
8. Record the packed cell volumes (PCVs) of the samples and discard the supernatants.
9. Resuspend the cells in 2 ml of distilled water and centrifuge the samples at 350 g for 2 min, and discard the wash.
10. Resuspend the cells in 1 ml of distilled water and store the samples at -70°C until all the samples are collected.
11. Bring the volume of the samples up to 3.5 ml with water and homogenize them with a Brinkmann polytron at high speed for 20 min.
12. Centrifuge the samples at 1400 g for 20 min.
13. Dilute the supernatant 1 in.1000 or 1 in 10000 with distilled water.
14. Measure the fluorescence of the diluted sample at 516 nm with an excitation of 488 nm.
15. Calculate the fluorescence intensity per volume of packed cells to determine relative viabilities of the cultures.

III. Regeneration of plants from toxin/filterate mediated selected cell lines

The applicability of *in vitro* selection to plant improvement is dependent on the expression of totipotency by the undifferentiated selected cells. However, a common limitation of long-term maintenance of plant cultures is the loss of the regenerative capacity of the material over time in culture. If the objective of the *in vitro* selection procedure is to obtain resistant plants then a scheme with alternating cycles of selection and regeneration may be the best approach (12).

An important step in the use of *in vitro* selection for plant improvement is to test the relationship between resistance of the plant cultures to the culture filtrate and the resistance of the plants regenerated from the cultures to the fungus. This is necessary because resistance to a fungal toxin at the cellular level may or may not be related to whole plant resistance to the fungus.

Equipment and reagents

* 12-day-old alfalfa cultures
* Plastic trays and transparent covers
* Nylon screens (500, 224, and 63µm)
* Pots
* Petri dishes containing Bi02y media
* Planting mix (Turface)
* Peat pellets.

METHOD :

1. Filter 10 ml of the filtrate-treated and control cultures, separately, through the 500µm mesh and discard the clumps that are collected.
2. Collect the proembryogenic masses from the filtered cultures by passing them through a 224µm mesh.
3. Transfer the proembryogenic cells on to a fine nylon screen (63µm mesh) and place the screens plus the cells into a Petri dish containing solid BiO2y medium.
4. Incubate the dishes in a growth cabinet set at 25°C and a 16 h light, 8 h dark photoperiod (90 $e.m^{-2}.s^{-1}$) for three to four weeks to allow the embryos to develop.
5. Transfer approximately 20 well-formed embyos to dishes with fresh BiO2y media.
6. Incubate the plates at 4°C in the dark for two weeks and then transfer them back to a growth cabinet set at 25°C and a 16 h light, 8 h dark photoperiod (90 $E.m\ -E^{2}.S^{-1}$) for three to four weeks to allow the embroys to develop shoots and roots.
7. Transfer plantlets formed from the embryos to moist peat pellets and place into trays covered with transparent plastic lids.
8. Maintain plantlets in a growth room or glass house (23/16°C and 16/8 h light (400-500 $E.m^{-2}\ s^{-1}$)/dark) with the lids until plantlets are acclimated to growth conditions out of culture.
9. When plants are approximately 4-6 cm high transplant them to plastic pots containing a planning mix.

IV. Testing of disease resistance of regenerated plants

Equipment and reagents

* *Verticillium* culture filtrate
* Alfalfa plantlets regenerated from culture
* Test-tubes
* Sterilized beaker

METHOD :

1. Prepare four stem cuttings per plant from the tops of actively growing plants so that each cutting consists of the appex plus three to four nodes.
2. Place the cuttings into a beaker containing 50% *Verticillium* culture filtrate plus 50% sterillized water for 24 h in a growth cabinet set to 18°C with 16 h illumination (400-500 $E.m^{-2}.s^{-1}$).
3. Transfer the cuttings into individual test-tubes containing sterilized water and seal the space between the steam and the top of the test-tube with Parafilm.
4. Maintain the cuttings in a growth cabinet set at 18°C with 16 h illumination (400-500 $E.m^{-2}.s^{-1}$) for seven days and rank the disease severity according to a scale from 1-5 described by Ireland and Leath (11)

   Where :

   1 = green leaves

   2 = general chlorosis of the leaflets

   3 = leaf necrosis with or without chlorosis and leaf curl on less than 50% of the leaves.

   4 = leaf necrosis and curl on more than 50% of the leaves

   5 = Wilted steam with chlorosis, necrosis, and leaf curl

   The cuttings with ratings of 1 or 2 are considered to be resistant, whereas, those with ratings of 3-5 are considered to be susceptible. The disease severity index of each plant is calculated from the mean of the four cuttings.

REFERENCES

1. Atanassov, A. and Brown, D. (1984) Plant Cell Tiss. Organ Cult., 3, 149.
2. Brazolot, J.K. Yu, F.K. and Paul, K.R. 1995. In Plant Cell culture-A Practical. Approch (Dixon, R.A. and Gonzales, R.A. eds.) Oxford Univ. Press. Oxford, pp 87.

3. Dische, Z. (1962) In : Methods in Carbohydrate Chemistry (ed. R.L. Whistler and M.L. Wolform), Academic Press, New York. pp. 4878-94.
4. Dixon, R. A and Lamb, C.J. (1990) Annu.Rev. Plant Physiol. Plant Mol. Biol., 41, 339.
5. Evans, D.A. (1989) TIG, 5, 46.
6. Frame, B., Yu, K.F., Christic, B.R and Pauls, K.P. (1991) Physiol Mol, Plant Patho., 39, 325.
7. Gamborg, O.L. Miller, R.A. and Ojma, K. (1968) Exp. Cell Res., 50, 151.
8. Gross, D.C. (1991) Annu. Rev. Phytopathol., 29, 247.
9. Ireland, K.F. and Leath, K.T. (1987) Plant Dis., 71,900.
10. Jones, P.W. (1990). In : Plant Cell Line Selection (Ed. P.J. Dix), VCH-Weinheim. pp. 113-49.
11. Negutiu, I (1990) In : Plant Cell Line Selection (Ed. P.J. Dix), VCH, Weinheim pp., 19-38
12. Ryan, C.A and Farmer, E.E. (1991). Annu. Rev. Plant Physiol, Plant Mol. Biol., 42, 651.

13

CRYOPRESERVATION FOR GERMPLASM CONSERVATION

Cryopreservation is the non-lethal storage of biological tissues at ultra-low temperatures. It is now possible to combine several cryopreservation techniques and apply them to a wide-range of plant systems. Cryopreservation comprises many steps, of which freezing is only one; successful recovery is dependent on the combined effects of cryogenics and pre and post-freeze treatments. Tissues must be selected from healthy plants and, in the case of *in vitro* material, culture parameters should be optimized before cryopreservation. Uniform suspensions composed of small groups of cytoplasmic, meristemtic cells are more suitable than vacuolated cells which have a high water content (1). The duration of pre-freeze culture and light regimes may affect shoot-tip recovery (2,3). Similarly, developmental and physiological factors are important determinants of recovery in embryo freezing. Pre growth treatment in combination with other cryoprotective treatments have the capacity to improve recovery. Pre-growth may also involve cold hardening treatment of the application of additives, known to enhance plant stress tolerance (e.g. proline, trehalose, abscisic acid.) Partial tissue dehydration can be achieved by the application of osmotically active compounds. The addition of relatively low (e.g. 1-5 %) concentrations of dimethylsulfoxide (DMSO) during pre-growth often improves shoot-tip recovery.

1. CRYOPROTECTIVE TREATMENTS

A. Chemical cryoprotection

Cryoprotectants are thought to exert their protective effects through colligative action which reduces potential toxic effect of freeze-induced solute concentration, the protective action of non-penetrating cryoprotectants is largely by osmotic dehydration. Other cryoprotective properties include the ability to alter biochemical and structural properties of membranes and thereby enhance freeze tolerance.

B. Vitrification

Avoidance of ice formation in biological tissues exposed to low and ultra-low temperatures reduces damage. This can be achieved through vitrification, a process in which ice formation cannot take place because the aqueous solution is too concentrated to permit ice crystal nucleation. In cryopreservation, vitrification usually involves the application of a highly concentrated solution of osmotically active compounds, many of which are used as 'traditional' cryoprotectants or pre-growth additives at lower concentrations. Care must be taken in the application of vitrification cocktails which become toxic on prolonged exposure. Further more, damaging osmotic affects may also occure on thier removal.

C. Cryotropective Dehydration

If cells are sufficiently dehydrated they may be able to withstand immersion in liquid nitrogen without the further application of traditional cryoprotectant mixtures. Dehydration can be achieved by growth in the presence of high concentrations of osmotically active compounds (sugars, polyols) and/or air desiccation in a sterile flow cabinet or over silica gel. Dehydration reduces the amount of water available for ice formation. If sufficiently dehydrated, cellular solute concentrations are high enough to promote vitrification.

D. Encapsulation

Encapsulation of tissues in calcium alginate beads which are pre-grown liquid culture media containing high concentrations of sucrose. The beads are transferred to a sterile air flow in a laminar cabinet and desiccated further. After these treatments the tissues are able to withstand exposure to liquid nitrogen without the application of chemical cryoprotectants (4).

E. Freezing and Storage

The means by which tissues are transferred from growth temperatures to ultra-low temperatures can be an important factor in post-freeze survival. Samples may be directly plugged into liquid nitrogen (rapid freezing), or progressively chilled at a controlled rate to an intermediate temperature. On reaching this temperature they may be transferred directly to liquid nitrogen. Samples are maintained in storage dewars and levels of liquid nitrogen must be regularly monitored and replenished as required.

F. Thawing

Thawing is particularly critical in vitrified tissues as ice crystallization can occure during re-warming. In conventional freezing methods, thawing is usually achieved by plunging the cryovial into sterile water maintained at 40-50°C. Once all the ice has melted, the samples are removed. Tissues which have been frozen by encapsulation and/or dehydration are frequently thawed at ambient temperatures. Vitrified and excessively dehydrated tissues may require a sequential lowering of media osmotica during thawing and early recovery.

G. Recovery

Post-thaw recovery comprises several stages : assessments of viability and cell division, morphological changes (leaf expansion, callusing, embryo development), regeneration of shoots and plants, and resumption of metabolite production. In the case of organized structures, recovery is frequently induced by the application of hormones.

2. CRYOPRESERVATION EQUIPMENTS

Simple freezing chambers with a cooled solvent system. However, a typical, programmable freezer comprises a liquid nitrogen cooled sample chamber, to which nitrogen is pumped from an accessory dewar or storage tank. Cooling rates, holding items, and terminal transfer temperatures are programmed.

You will need the following equipments.

1. a reliable source of liquid nitrogen.
2. safety equipment (gloves, apron, face shield, pumps for dispensing liquid nitrogen from and to large storage dewar's, trolleys for the transport of dewar's).
3. small (1-2 litre) liquid nitrogen resistant dewar(s).
4. dewar(s) for the routine storage of liquid nitrogen
5. dewar(s) of the long-term storage of specimens-these may be equipped with an alarm system which is activated if the liquid nitrogen level falls below a critical level.

6. cryovials, straws, boxes, canes, racks.
7. a domestic refrigerator (-20°C)
8. a programmable freezer with dewar and pump
9. a water-bath for thawing at 40-50°C

3. PROTOCOLS FOR METHOD DEVELOPMENT

The following protocol offers an example which may be used for a range of different tissues. Treatments can also include cold-hardening strategies which are frequently linked to seasonal factors such as dormancy (4). Because of the specific nature of cold-hardening treatments, it is not possible to present a general methodology (5).

Pre-growth

1. Prepare standard liquid or semi-solid medium and add one or more of mannitol, sucrose, proline, sorbitol, trehalose, or glucose in a range of 0.3-0.5 M.
2. Culture tissues with the additives for one to five days.
3. At daily intervals, return tissues to standard growth.
4. Determine survivial for each treatment.

   Note: In developing a pre-growth method for shoot-tips, it is essential to standardize the donor material as variation in response to freezing is a major problem in shoot-tip cryopreservation. Dissection injury can often be a significant deteriminant of post-freeze survival, particulary if the shoot-tips are forzen directly after excision (6).

Shoot-tip dissection

5. Select apical shoots at the same development age or cultures of the same subculture age. For rapidly growing cultures, nodal sections can be cultured for five to ten days (dependent on species) and the developing axillary bud used for cryopreservation.
6. Remove the shoot-tip and place on sterile filter papers under a binocular microscope. With the aid of two fine-gauge hypodermic needles (fitted to a syringe), gently disect away any fully expanded leaves and cut the shoot-tip to a size of 0.3-0.5 mm (dependent on species). Retain three or four non-expanded leaf primordia on each shoot-tip. It may be necessary to optimize both shoot-tip size and the period required for recovery.
7. Immediately after dissection, place the shoot-tip onto a filter paper bridge moistened with standard liquid culture medium. The

application of 0.5-5% (v/v) DMSO at this stage may enhance survival.

8. Pre-grow for one to three days before cryopreservation.

(a) Cryoprotection of cell and callus cultures (7) :

9. Prepare one of the following cryoprotectants in standard liquid culture medium:

   * 1 M DMSO + 1 M glycerol + 2 M L-proline
   * 1 M DMSO + 1 M glycerol + 2 M sucrose

10. Divide the cryoprotectant into two aliquots, chill one aliquot in iced water and maintain the second at 25°C.
11. Remove excess liquid from replicate flasks of cell suspension clutures leaving a known volume of culture.
12. Add chilled or 25°C cryoprotectant mixture to replicate cultures in the volume ratio of 1:1, by either direct or gradual addition (2-3 ml aliquots over 30 min). Maintain the chilled replicate on ice throughout the incubation time.
13. Cryoprotect for 1 h (inclusive of addition time).
14. Empty flask contents onto filter papers, assess cell viability in few cells. Transfer remaining cells to semi-solid cultures medium and assess re-growth, Repeat the process with the second cryoprotectant and determine the optimum procedure.

(b) Cryoprotection of meristem structure (8).

15. Prepare cryoprotectant solutions of DMSO in standard liquid medium in the rangc 5-15% (v/v)
16. Add 10-15 meristem structures (Shoot or root-tips) to 0.5-0.75 ml of cryoprotectant medium (at 25°C).
17. Cryoprotect for 15 min to 1h.
18. Remove meristem structures, blot dry on filter paper, transfer to recovery medium, and assess re-growth in comparison to uncryoprotected controls, Select the least damging treatment for cryopreservation (9, 10).

(c) Cryoprotective Chemical Dehydration (11, 12, 13)

19. Prepare standard liquid or solid medium and add one of the following: mannitol, sucrose, glucose, or sorbitol, in the concentration range 0.50-1.5M.
20. Determine tissue fresh weight then transfer to media.

21. Dehydrate tissues for short (2-24h) and long-term periods (one to seven days).
22. Determine fresh weights and return to standard medium. Determine dry weights of replicate samples.
23. Correlate maximum water loss with minimum mortality. Select this treatment for prefreeze dehydration (14).

(d) Cryoprotective Air Dehydration

24. Determine tissue fresh weight.
25. Dehydrate tissues for several time intervals in the range 1-16 h by using one of the following methods :
    * Sterile air flow
    * Flash drying in a pressured stream of sterile air
    * drying over silica gel
26. Determine fresh weights of tissues and return to standard culture medium. Determine dry weights of replicate samples.
27. Correlate maximum water loss with minimum mortality.

(e) Encapsulation and Dehydration

28. Prepare :
    * Calcium-free standard liquid medium 3% (w/v) sodium alginate (2% viscosity Kelp Na^+ salt)
    * Standard liquid medium containing 100 mM $CaCl_2$
29. Dissect shoot-tips
30. Add shoot-tips to a flask and add ~ 25 ml alginate solutions, then swirl until shoot-tips are completely submerged. Take care not to form air bubbles.
31. Place ~ 50 ml of 100 mM $CaCl_2$ medium into a 100 ml conical flask, using a 5ml 'Pipetteman' withdraw 3 ml of the alginate solution and, with gentle manipulation, take up serval shoot-tips at the same time.
32. Wipe away any excess alginate from the tip (failure to do so will result in asymmetric beads).
33. Place the pipette tip over the flask containing the calcium medium and dispense drops into the liquid; beads form as the alginate solidifies on contact with the Ca^{2+} ions. It is possible to encapsulate one to three shoot-tips per bead.
34. Pour off excess medium, pick out the beads with the shoot-tips,and allow the beads to polymerize for 30 min.

35. Divide the encapsulated shoot-tips into replicate samples, blot dry on filter paper, and transfer to the following ;
    * Standard semi-solid culture medium
    * Arange of cryoprotective dehydration media up in liquid culture medium (Step 19).
36. Dehydrate the beads in liquid culture for short (2-16 h) and long-term periods (one to five days)
37. At appropriate time intervals remove replicate samples from each treatment; place one replicate onto standard semi-solid medium.
38. Transfer the remaining replicates to open Petri dishes in a sterile laminar air flow, air dry 1-4 h, and each time interval transfer to standard semi-solid medium.
39. Determine levels of survival for each stage and select the treatment which gives maximum survival for maximum dehydration.

4. VITRIFICATION

Vitrification solutions are potentially very toxic, and critical factors include the cold acclimation of tissues before vitrification, the temperature of application of vitrification solution, and the duration of exposure. There have been several successful reports of vitrification (8,15, 16). Sakai et. al. (3) have developed the 'PVS2' vitrification solution which has proved applicable to a range of systems (14, 32, 17).

1. Prepare the following solutions :
    * 'PVS2' stock solution : 30% (w/v) glycerol + 15% (w/v) ethylene glycol + 15% (w/v) DMSO in standard liquid medium containing 0.15 M sucrose.
    * 60% (v/v) 'PVS2' solution in standard liquid medium containing 0.4 M sucrose.
    * 'unloading' solution : 1.2 M sucrose in standard liquid medium
2. Add 1.0 ml 25°C (v/v) 'PVS2' solution to the tissues.
3. Incubate at 25°C for 5-20 min.
4. Remove 60% solutions, replace with 1ml chilled 100% 'PVS2' and incubate tissues in an iced water-bath for 3-10 min.
5. Transfer to 25°C and remove excess 'PVS2' solution.
6. Add 1.0 ml unloading solution and change three times. Incubate in unloading solution for 5-30 min.

7. Transfer to filter papers overlaid on standard semi-solid culture medium. On the following day transfer to fresh medium. Evaluate survival for each treatment.

5. FREEZING

In the case of cell suspension cultures, it is often possible to apply the method of Winthers and King (1). However, shoot-tips often have highly species specific controlled cooling requirements (18). Under given protocol offers one possible regime for optimizing controlled cooling (19).

1. Transfer cryoprotected tissues to a number of replicate cryovials.
2. Set the starting temperature of the freezer to 0°C or 25°C depending on the temperature of cryoprotection (see earlier protocol)
3. Programme one of the following cooling rates:– -5, -1.0, -0.5, -0.25, -0.1°C/min.
4. Cool to intermediate transfer temperatures : -0, -10, -15, -20, -25, -30, -40°C, and remove vial (s) as appropriate.
5. Thaw at 45°C until the ice has melted and transfer the tissues to standard semi-soild medium.
6. Determine optimum transfer temperature. Repeat under optimum conditions, but this time transfer the tissues to liquid nitrogen.
7. Assess recovery, if the programme was unsuccessful, repeat steps 1 to 6 using different cooling rates. Use 'control' samples which have not been exposed to liquid nitrogen to intemediate survival.
8. If tissues are still not able to survive exposure to liquid nitrogen test, use the following variables:
   * 'Seeding' ice crystallization (See protocol for cryopresevation of shoot tips.)
   * holding for 20-40 min at the terminal temperature.
   * stepwise cooling (e.g. -1.0°C/min to -15°C, followed by -0.5° C/min to -40°C).

6. THAWING AND RECOVERY

There are few variables at the thawing stages of method development. Recovery regimes are species specific and influenced by tissue culture methods.

7. PROTOCOLS FOR SPECIFIC APPLICATIONS

(i) Protocols for cryopreservation of cell cultures (1, 20, 21).

1. Select cell culture, taking into consideration pre-growth factors.
2. Remove excess culture medium to a known volume and add the ice-chilled cryoprotectant (1.0 M DMSO + 1.0 M glycerol + 2.0 M sucrose) in a volume ratio of 1:1.
3. Mix the cells thoroughly, cryoprotect on ice for 1h, and transfer 1-1.75 ml aliquots of the cryoprotectant- cell mixture to 2.0 ml cryovials, Load the vials onto a cane and transfer to a programmable freezer, pre-chilled to 0°C.
4. Use the following conditions to initiate the chilling programme.
   * Chilling rate - 1°C/min
   * Terminal temperature - 35°C
   * Holding time 30 min
5. On completion, quickly transfer the canes to a long-term storage dewar containing liquid nitrogen.
6. Transfer vials to a 45°C water-bath, and, once the ice has melted, pour their contents onto several layers of 5.0 cm filter papers in a 9.0 cm petri dish.
7. Transfer the cells and the top few layers of filter papers to semi-solid culture medium.
8. Check early post- thaw viability.
9. After one day, transfer the uppermost layers of filter papers/cells to fresh semi-solid medium and continue to transfer on a daily basis (one to four days.) Transfer cells to fresh medium without filters (five to six days).
10. Initiate liquid suspensions when re-growth is sufficient to permit a normal subculture inoculation.

(ii) Protocols for Cryopreservation of shoot-tips (22,23)

1. Dissect shoot-tips and pre-grow for 24h on filter paper bridges with liquid culture medium.
2. Transfer batches of 15-20 shoot-tips to cryovials containing 0.5-1.0 ml of 10% (v/v) DMSO.

3. Cryoprotect for 1 h at 25°C
4. Transfer to a programmable freezer pre-chilled to -5°C and chill for 10 min.
5. Induce ice ('seeding') by mechanical agitation or by touching the outside of the vial with liquid nitrogen-chilled forceps.
6. Chill for a further 10 min at -5°C.
7. Cool at a rate of -0.4°C/min to a terminal temperature of -35°C and transfer to liquid nitrogen.
8. Thaw at 45° C.
9. Empty the contents onto filter papers to remove cryoprotectant.
10. Transfer shoot-tips to standard semi-solid medium containing the following hormone combination:
    * 0.2 mg/litre gibberellic acid (GA_3)
    * 0.5 mg/litre indole acetic acid
    * 0.5 mg/litre zeatin (mixed isomers)
11. When shoot regeneration has occurred transfer to standard culture medium.

(iii) Protocol for cryopreservation of immature and mature zygotic embryos using combined cryoprotective dehydration.

A. Immature embryos (24)

1. Place embryos onto standard semi-solid medium containing 600g/litre glucose (replacing standard sugars) and 15% (v/v) glycerol.
2. Dehydrate for 4h, transfer to cryovials, and plunge directly into liquid nitrogen.
3. Thaw for 30 sec at 40° C and transfer to standard medium for recovery.

B. Mature embryos (25)

1. Place embroys in open Petri dishes (without medium) and desiccate in a sterile air flow for 4h.
2. Transfer to medium containing 600g/liter glucose and 15% (v/v) glycerol, dehydrate for 15-24 h (Cultivar dependent.)
3. Freeze, thaw, and recover as for immature embryos.

(iv) Protocols for the cryopreservation of somatic embryos using cryoprotective dehydration (28)

1. Culture embryonic masses on standard semi-solid medium containing 0.3 M sucrose for two months.
2. Transfer to standard semi-solid medium containing 0.75 M sucrose for seven days.
3. Dissect out embryonic tissue, transfer to cryovials, and plunge directly into liquid nitrogen.
4. Thaw at 45°C then transfer to standard semi-soild medium containing 0.3 M sucrose. After two weeks transfer to standard medium containing 0.1 M sucrose.

(v) Protocol for cryopreservation of shoot-tips and somatic embryos by encapsulation and dehydration

1. Dissect shoot-tips.
2. Encapsulate shoot-tips or torpedo stage somatic embryos as described earlier.
3. Transfer beads to standard liquid medium containing 0.75 sucrose (for shoot-tips) or .3 M sucrose (for embryos). Place on a shaker and dehydrate for 72 h (shoot -tips) or 18h (embryos).
4. Transfer beads to filter papers, blot, dry, remove filters, and desiccate in a sterile air flow for 4 h.
5. Transfer the dried beads to a cryovial and plunge directly in liquid nitrogen.
6. Thaw at ambient temperature (~25°C) and transfer the beads to semi-solid recovery medium. If necessary excise tissues from the beads to aid recovery (26, 27).

(vi) Post-thaw viability, recovery, and stability.

Vital staining techniques like FDA and TTC are useful in determining early postthaw survival. However, they are not always predictive of long-term viability and recovery and their use is coutioned as the sole means of assessing survival. Ideally, tests should be performed at different stages after thawing. The only definitive indication of recovery is re-growth. Retention of genetic fidelity in plants recovered from cryopreservation is vital. Molecular probes for ribosomal DNA may be used to study restriction fragment length polymorphisms in plants recovered from cryopreservation. This offers a potentially important method of assessing genetic stability (29). Similarly, the use of flow cytometry to assess post freeze ploidy stability may prove a repaid and valuable technique (30, 31).

REFERENCES

1. Withers, L.A. and King, P.J. (1980). Cryo-Lett., 1, 213.
2. Benson, E.E., Harding, K., and Smith, H. (1989) Cryo-Lett. 10, 323.
3. Harding, K., Benson, E.E. and Smith, H. (1991) Cryo-Lett., 12, 17.
4. Tyler, N., Stushnoff, C., and Gusta, L.V. (1988) Can, J. Plant Sci., 68, 1169.
5. Goldner, E.M. Seitz, U., and Reinhard. E (1991). Plant Cell Tiss. Organ Cult., 24, 19.
6. Henshaw, G.G. O'Hara, J.F. and Stamp, J.A,. (1985). In: Cryopreservation of Plant Cells and Organs (ed. K.K. Kartha), p. 159-70. CRC Press, Boca Raton, Florida.
7. Finkle, B.J. Zavala, M.E and Ulrich, J.M. (1985). In : Cryopreservation of Plant Cells and Organs (ed. K.K. Kartha), pp. 76-113. CRC Press, Boca Raton, Florida.
8. Langis, R. and Steponkus, P.L. (1990). Plant Physiol., 92, 666.
9. Benson, E.E. and Hamill, J.D. (1991). Plant Cell Tiss. Organ Cult., 24, 161
10. Demeulemeester, M.A.C. Panis, B.J. and De proft, M.P. (1992), Cryo-Lett. 13, 165.
11. Engelmann, F. (1990) C.R. Acad. Sci. Paris, SIII 311, 679.
12. Engelmann, F., Duval, Y., and Dereuddre, J. (1985) C.R. Acad. Sci Paris, S. III, 301, 111.
13. Dumet, D., Engelmann, F., Chabrillange, N. and Duval, Y. (1993). Plant Cell Rep. 12, 352.
14. Uragami, A. Sakai, A., and Nagai, M (1990) Plant Cell Rep., 9, 328.
15. Towill, L.E. (1990) Plant Cell Rep,. 9, 178.
16. Langis, R.L. and Steponkus, P.L (1991) Cryo-Lett., 12, 107.
17. Yamada, T., Sakai. A., Matsumura, T., and Higuchi, S. (1991), Plant Sci., 78,81.
18. Kartha, K.K. (1985). In cryopreservation of Plant Cells and Organs (ed. K.K. Katha), pp. 115, 34, CRC Press, Boca Raton, Florida.
19. Heszky, L.E., Jekkel, Z., and Abdel-Hamid, A, (1990). Plant Cell Tiss. Organ Cult., 21, 217.
20. Meijer, E,G.M. Van Iren, F., Scrinjnekmakers, E., Hensgens, L.A., van Zijderveld, M., and Schilperoot, R.A. (1991) Plant Cell Rep. 10, 171.
21. Panis, B.J. Withers, L.A., and De Langhe, E.A.L. (1990) Cryo-Lett., 11, 337.
22. Towill L.E. (1983) Cryobiology, 20, 567.
23. Towill L.E (1984). Cryobiology, 5, 319.

24. Marin, M.L. and Duran-Vila, N. (1988). Plant Cell Tiss, Organ Cult., 14, 51.
25. Withers, L.A. (1979). Plant Physiol., 63, 460.
26. Niino, T., Sakai, A. and Yakuwa, H. (1992). Cryo-Lett., 12, 135.
27. Widholom, J.M. (1972) Stdin Technol., 47, 189.
28. Steponkus, P.L. and Lamphear, F.O. (1967) . Plant Physiol, 42, 143.
29. Harding, K. (1991). Euphytica, 55, 141
30. Ward, A.C.W., Benson, E.E., Blackhall, N.W. Cooper-Bland, S., Poweell, W., Power J.B. and Davey, M.R. (1993). Cryo-Lett., 14, 145.
31. Bensin, EE.1994. In Plant Cell Culture - A Practical Approach 2nd Ed. (Dixon, R.A. and Gonzales, R.A. eds) Oxford Univ. Press Oxford P.P. 47-167.
32. Sakai, A., Kobayashi, S. and Oiyanma, I. (1990). Plant Cell Rep., 9, 30.

14

Determination of Endogenous Content of Cytokinins in Plant Material

Experimental Material

The Method is applicable for every kind of plant material (1).

Extraction and purification Chemicals and buffers

80% methanol;

0.04.M ammonium acetate buffer, pH = 6.5;

DEAE - cellulose;

^{3}H-ZR (15000 dpm/50µl);

Equipments

1. mortar with pestle or homogenizer;
2. centrifuge, 15000 rpm;
3. centrifuge tubes;
4. flasks for evaporation, 100ml and 20ml;
5. rotary evaporator;
6. syringes - 5ml, 10ml, 50ml.
7. C_{18} SepPak cartridges

PROCEDURE

1. Freeze and powder 1g plant material in a mortar with liquid nitrogen. Add 10 ml 80% methanol, homogenize the powder and transfer it into a 15ml centrifuge tube.
2. Add to the sample 50 µl ^{3}H-ZR as a radiactive internal standard for recovery.
3. Incubate the sample in freezer at -20°C for 1 hour.
4. Centrifuge at 15000 rpm for 20 min at 4°C.

5. Take out the supernatant and pass it through a C_{18} SepPak cartridge (pre rinsed with 10 ml technical ethanol and 20 ml distilled water).
6. Evaporate the methanol with rotary evaporator at 35-40°C.
7. Add to the water residue 25 ml 0.04 M ammonium acetate buffer, pH=6.5
8. Pass the sample through a DEAE-cellulose column connected to a C_{18} SepPak cartridge (Pre-rinsed as above) in 0.04 M ammonium acetate buffer, pH 6.5
9. Wash the column with 50 ml of the same buffer.
10. Disconnect the cartridge, wash with 10 ml dH_2O and elute the cytokinins from it with 10 ml 80% methanol into 20 ml flask for evaporation.
11. Evaporate to dryness with rotary evaporator, 35-40°C.
12. Dissolve the sample in 2 × 500μl methanol by vortexing and sonicating and transfer to an Eppendorf tube for lyophilization.
13. Dissolve in 30μl 50% methanol for HPLC separation.

HPLC separation Chemical and buffers

Phase A

80% methanol, UV grade in TEA*

Phase B

20% methanol, UV grade in TEA.

* TEA-40 mM acetic acid, adjusted pH=3.35 with triethylamine.

1μg/ml standard samples of each: zeatin (Z), zeatin riboside (ZR), isopentenyl-adenine (2iP), isopentenyl-adenosine (2iPA), dihydrozeatin (dHZ) and dihydrozeatin-riboside (dHZR).

Equipments

1. Liquid chromatograph;
2. C_{18} ODS column;
3. UV- detector, 268 nm;
4. Hamilton pipette, 10μl;
5. Polystirol tubes 5 × 1.2 cm.

Conditions

Gradient	0 - 3 min	10% A - 10 A
	3 - 15 min	10% A - 70% A
	15 - 25 min	70% A - 70% A
	25 - 27 min	70% A - 10% A
Flow rate	0.7 ml/min	
Temperature	25°C.	

PROCEDURE

1. Take out 100µl of each standard solution for reparing a standard mixture.
2. Inject 10µl of the standard mixture and check the retention time of each fraction. Repeat this step 2-3 times.
3. Inject 10µl of the sample and collect the fractions, corresponding to the retention times of Z, ZR, 2iP, 2iPA, dHZ and dHZR in polystirol tubes.
4. Evaporate the fractions to dryness.
5. Dissolve in 1 ml buffer for ELISA.
6. Take out 150 µl of the sample to measure the recovery

ELISA assay

Chemicals and buffers

Coating buffer (0.05 M Na_2CO_3/$NaHCO_3$, pH=9.6);

PBS buffer (phosphate buffered saline, pH=7.4)

PBST buffers (PBS + 0.05% Tween 20);

Substrate buffer (Coating buffer + 0.5 mM $MgCl_2$);

Polyclonal antibodies against ZR, 2iPA and dHZR;

Enzyme conjugates : ZR, 2iPA and dHZR coupled to alkaline phosphatase; Substrate for alkaline phosphatase - p-nitrophenyl phosphate;

Standard solution of Z, ZR, 2iP, 2iPA, dHZ and dHZR in range of 0.5-1000 pmol/150µl in PBS;

0.25% Bovine serum albumin (BSA) in PBS;

Equipments

1. ELISA microtitre plates;
2. Titertek;

3. Fine pipette : 5 - 20µl, 20 - 200µl and 50 - 1000µl;
4. Thermostate - 37°C;
5. ELISA-reader;
6. Liquid scintillation analyzer.

PROCEDURE

1. Fill each well of the ELISA plate with 200pl antibody in Coating buffer (5 µg/ml);
2. Incubate overnight at 4°C;
3. Rinse the wells with PBST (5 times);
4. Fill the wells with 250µl BSA, incubate for 20 min at room temperature and wash 5 times with PBS;
5. Add to each well 150µl standard or sample in several dilutions, both in PBS;
6. Add in separate wells PBS or large excess of standard (1000 pmol/150µl) for determination of maximal antibody binding (B_0) and minimal antibody binding (B_u);
7. Incubate at 37° C for 30 min.
8. Add to each well 50 µl enzyme conjugate;
9. Incubate for 2.5 hours at 37° C;
10. Wash the plates with PBS (5 times);
11. Add the substrate (200 µl 1 mg/ml p-nitrophenyl phosphate in substrate buffer);
12. Incubate for 60 min at 37°C.
13. Measure the optical densities at 405 nm by ELISA-reader;
14. Built mean standard curves for each cytokinin using B/B_0 vs lg [pM/150 µl]. For determination of B/N_0 use the formula :

    B - the absorbance at 405 nm in the presence of a known amount of standard or unknown sample;

    B_u - the absorbance at 405 nm in the presence of excess (1000 pmol) of the standard;

    B_0 - the absorbance in absence of standard;

    Determine the sample concentration using the standard curves and B/B_0 value; Measure initial and final radioactivity of the sample and calculate the recovery using the equation :

$$\frac{A}{A^*} = \frac{a}{a^*}$$

Where

A - the cytokinin amount at the beginning of the analysis;

a - the cytokinin amount, measured at the end of the analysis;

A* - the radioactive standard at the beginning of the analysis;

a* - the radioactive standard at the end of the analysis.

REFERENCES

1 Anon. 1995. Training Manual of ICGEB Practical Course on Gene Transfer for Plant Breeding Improvement. Institute of Genetic Engineering, Kostinbrod (Bulgaria).

15

IDENTIFICATION OF VIRUSES AND/OR NUCLEOCAPSID PROTEIN BY ELISA

The levels of virus infection or the content of coat (nucleocapsid) protein in tansgenic plants, can be measured by various methods. One of the most widely used one is ELISA. It is applied in various modifications and has the following advantages :

- possibility for simultanious analysis of a large number of samples
- Extreme sensitivity and reliability
- Economy in terms of use of antiresa
- Rapidity
- It gives not only qualitative/quantitative data, but also such on the proteins quantity, regardless virus morphology and concentration (1).

Before ELISA to be applied, one should have obtained quatity antisera. This can be achieved if the virus is propagated on the indicator plant, further purified following a special methodology and then injected into a hot blood animal (usually rabbits). Thus, antibodies are produced against the virus or nucleocapsid protein, which antibodies would interact with the same antigene. The lgG (immunoglobuline G) fraction is then isolated from the antisera containing these antibodies and part of the fraction is marked with an enzyme. The more widely used enzymes are alkaline phosphatase and peroxidase, which possess specific substrats. Currently, gamma-glubolines (lgG) and conjugates are commercially supplied by many companies.

Materials

1. Microtitre plates
2. Adjustable hand pipets with disposable tips
3. Themostate and refrigerator
4. Photometer

Buffers and solutions

1. Buffer A, coating buffer carbonate, pH 9.6 for dilution of the first antibody

Na_2CO_3	1.59g
$NaHCO_3$	2.93g
NaN_3	0.2g

Make it to 1L with distilled water

2. Buffer B, phosphate buffered along pH/4 (PB).

NaCl	8.0g
KH_2PO_4	0.2g
$Na_2HPO_4\ 12H_2O$	2.9g
KCl	0.2g
(0.2g NaN_3)	
Tween 20	0.05%

Make it to 1L with distilled water

3. Substrate buffer C, pH 9.6 for alkaline phosphatase conjugate, diethanolamine 97 ml

NaN_3	0.2 g

are mixed in 800ml distilled water, pH is adjusted with HCl and solution is diluted to 1L.

4. Substrate buffer C_1, pH 5-Na-citrate buffer for peroxidase conjugate

   0.1 M citric acid

   0.2 M $Na_2\ HPO_4.12H_2O$

   0.002% H_2O_2

5. Phosphate buffer for infection of the plants (buffer D)

   13.6g KH_2PO_4 in 1L dH_2O

   35.8g $Na_2HPO_4.12H_2O$ in 1L dH_2O

   for pH 7.0, 30 ml of the first buffer and 70ml of second buffer + 2% PVP_{40}, 0.5% Na_2SO_3 and 0.2% ascorbic acid.

MEASURING THE NUCLEOCAPSID PROTEIN WITH TAS-ELISA

1. The ELISA plates already done have to be washed with distilled water and dried. Put into appropriate wells (without 1A) 50-200µl polyclonal antibody- lgG in concentration 1 µg/ml dissolved in buffer A (coating buffer). Cover the plates and incubate for 240 minutes at 35°C.

2. Extract leaves and label them appropriately (usually 50mg leaf tissue). Grind leaf tissue with 5 ml buffer B (1:100) which contains 0.5% Na_2SO_3 samples are loaded in 2-4 repeats and incubated for approximately 16 hours at 4°C.
3. The monoclonal antibody is diluted 1:2000 in PBS-T, containing 2% polyvinylpyrrolidone (PVP_{40}) and 0.2% egg albumin. The wells are covered (without 1A). and plates are incubated for 2 hours at 35°C.
4. The labeled with enzyme antimouse IgG developed in rabbit (conjugate) dilute 1:5000 in PBS-T and containing 2% BSA covers all wells of the plaque without 1A and incubated for 2 hours at 35°C. Triple wash with PBS-T and incubated for 2 hours at 35°C.
5. Usually four substrats are used p-nitrophenylphosphat 0.6-1 mg/ml in diethanolamine buffer when worked with alkaline phosphatase. Substrate for peroxidase is o-phenylendiamin 4mg/10ml Na citrate buffer, pH 5 with H_2O_2 and final dilution 0.002%. Cover the whole plates with the substrate solution (buffers C and C_1).

The reactions develop from 10 to 30 minutes and they can be stopped with 3M NaOH (Phosphatase conjugate) or with 4M H_2SO_4 (Peroxidase conjugate).

The measurement is done photometrically at absorbanse 405 and 495 nm respectively.

INOCULATION OF THE TRANSGENIC PLANTS WITH TSWV

Leaves from a defined strain of the virus, propagated on indicator plants are extracted with buffer D. With the obtained innoculum 2-3 marked leaves from the plant which previously has been known to contain or not nucleocapsid protein. The plant had been numbered. The inoculation is done gently to avoid cracking the fibres. The incubation period is about ten days. In case no clear symptoms appear, a second inoculation and even a third one is made. This is required because of the easily inactivation of the rather instable TSWV. Finally on each of the series (control), nontransgenic plants are inoculated. The inoculation with purified virus preparation (with a determined concentration) is a better way for testing the resistance of transgenic plants against a given virus. However, it is still difficult to say whether a plant line is resistant to a virus. The processes of mechanical and/or vector transmission of TSWV depends of many factors, such as the light regime, humidity, temperature, the age of the plant to be inoculated.

The results obtained are summarized and, if required, can be statistically adjusted.

REFERENCES

1. Clark M.F., Adams, A.N. 1977. J. Gen. Virology 34, 457-483.
2. Hill A.S., 1984. Methods in plant virology. Blackwell Sci. Publ., 25-40.
3. Nordam D., 1973. Identification of plant viruses: methods and experiments, Genet. Agric, Publ. Docum., Wageningen.

16

ISOLATION AND CULTURE OF PLANT PROTOPLASTS

Protoplasts are the cells devoid of cell wall, offer unique opportunities for genetic manipulation at single cell level (Fig: 16.1). One can either introduce small amount of genetic material into protoplasts through electroporation, PEG-mediated uptake, liposome mediated transfer of DNA or bring about large scale transfer of genetic material through fusion of protoplasts from two different species or genera. Any type of tissue can be used for obtaining protoplasts. However, young seedling explants or leaves from *in vitro* shoot cultures give uniform, highly-viable and consistent yield of protoplasts. Protocols for isolating protoplasts from various crop species viz. sunflower, *Brassica*, Rice and tobacco are as follows :

1. PROTOPLAST ISOLATION AND CULTURE FROM HYPOCOTYLS OF SUNFLOWER

(*Helianthus annuus* L) (Punia, 1990; Punia *et. al.*, 1994; and Punia and Bohorova, 1996)

Media and solutions required.

(a) Murashige and Skoog medium (MS-medium)

Major Salts	**mg/1**
KNO_3	1900
NH_4NO_3	1650
KH_2PO_4	170
$MgSO_4.7H_2O$	370
$CaCl_2.2H_2O$	440
Minor Salts	
$FeSO_4,.7H_2O$	27.8

Na_2 EDTA.$2H_2O$	37.3
H_3BO_3	6.2
$MnSO_4.4H_2O$	22.3
ZnSO4.$7H_2O$	8.6
$CoCl_2.6H_2O$	0.025
$CuSO_4.5H_2O$	0.025
$Na_2MoO_4.2H_2O$	0.25
KI	0.83
Meso-inositol	100
Nicotinic acid	0.5
Pyridoxine HCl	0.5
Thiamine HCl	0.1
Glycine	2.0
Sucrose	30,000
Agar	8,000
pH	5.8

(b) Kao's medium for protoplast culture

NH_4NO_3,	600	KI	0.75
KNO_3	1900	H_3BO_3	3.0
$CaCl_2.2H_2O$	600	$MnSO_4.H_2O$	10.0
$MgSO_4.7H_2O$	300	$ZnSO_4.7H_2O$	2.0
KH_2PO_4	170	Na_2MoO_4, $2H_2O$	0.25
KCl	300	$CuSO_4.5H_2O$	0.025
Sequestrene 300Fe	28	$CoCl_2.6H_2O$	0.025
Glucose	72000	Mannose	125
Sucrose	125	Rhamnose	125
Fructose	125	Cellobiose	125
Ribose	125	Sorbitol	125
Xylose	125	Mannitol	125
Sodium Pyruvate	5	Malic Acid	10
Citric acid	10	Fumaric acid	10
Inositol	100	Folic acid	0.2
Thiamine HCl	10	Riboflavin	0.1
Ascorbic acid	1	p-amino benzoic acid	0.01

Nicotinamide	1	Vit. B_{12}	0.01
Pyridoxin HCl	1	Vit. A	0.005
Calcium pantothe	0.5	Vit. D_3	0.005
Choline Chloride	0.5	Biotin	0.005
Casein hydrolysate	100	Coconut water	10 ml
2, 4-D	1.0		
BAP	0.5		
NAA	0.1		

(pH 5.7)

(Filter sterilize).

(c) CPW 9M and CPW 13M Solutions

Component	Cnce (mg/l)
KH_2PO	27.2
KNO_3	101
$CaCl_2.2H_2O$	1480
$MgSO_4.7H_2O$	296
KI	0.16
$CuSO_4.5H_2O$	0.025
Mannitol	9% (90g) and 13% (130g) for CPW 9M & CPW 13M respectively.

(d) CPW 21S : All above components without mannitol, and add 21% Analar sucrose. All CPW's solutions i.e. CPW9M, CPW 13M and CPW 21S are autoclave sterilized.

(e) Enzyme Mixture

Enzymes	Concentration (%)		
	For hypocotyl explant of cultivated sunflower	For explant leaf of cultivated sunflower	For explant leaf of wild sunflower species
Rhozyme HP 150	2.0		
Meicelase	2.0		
Macerozyme	0.03		
Cellulase		1.0	2.0
Drislase		0.1	
Pectolyase		0.1	0.05

Enzymes are dissolved in CPW 9M or CPW 13M (for hypocotyl and leaf explants respectively) on magnetic stirrer and their pH is adjusted to 5.8. Enzyme mixture is centrifuged for 10 min. and their supernatant is first filtered through 8.0μ millipore sterilized filter and subsequently again filtered through 0.22μ sterilized millipore filter and is kept in sterilized bottles and stored at -20°C.

(f) Reagents :

Florecein diacetate (FDA)

DETERMINATION OF PROTOPLAST VIABILITY

Fluorescein diacetate (FDA) is often used to determine protoplast viability. FDA molecules are non-fluorescent, but are able to pass freely across the plasma membrane. Esterases within the cytoplasm of living cells cleave FDA to release the fluorescent compound fluorescein, which is unable to pass out through the plasma membrane of viable protoplasts. Mix 1 ml of a 5mg/ml stock solution of FDA with 10ml CPW salts solution to prepare a working dilution. Mix equal volumes of the working dilution of FDA and the protoplast suspension. After five minutes at 22°C, examine the protoplasts with a fluorescence microscope, using blue illumination. Viable protoplasts emit a green/yellow fluorescence, whereas, the non-viable protoplasts remain unstained.

COUNTING PROTOPLASTS USING A HAEMOCYTOMETER

1. Prepare protoplast suspension in a balance salt solution/culture medium containing approximately 2-5×10^5 protoplast/ml.

2. With the coverslip in place, use a pasteur pipette and transfer a small amount of protoplast suspension to both chambers of the haemocytometer by carefully touching the edge of the coverslip with the pipette tip and allowing each chamber to fill by capillary action. Do not overfill or underfill the chambers.

3. Starting with 1 chamber of the haemocytometer, count all the protoplasts in 1mm centre square and four 1 mm corner squares.

4. Repeat this procedure for chamber 2.

5. Each 1 mm square of the haemocytometer, with cover slip in place, represents a total volume of 0.1 mm^3 or 10^{-4} cm^3. Since 1 cm^3 is equivalent to 1ml, the protoplast concentration per ml will be determined as follows :

Protoplast per ml = the average count per square × dilution factor × 10^4

e.g. if the average court per square is 45 protoplasts then the total protoplasts/ml will be 45×10^4 or 4.5×10^5 protoplasts/ml

Hence, total protoplast in the tube containing 10 ml of original protoplast suspension will be:

$$4.5 \times 10 \times 10^5 = \quad 4.5 \times 10^6 \text{ protoplasts.}$$

Equipments and Glasswares required :

1. Laminar flow
2. Centrifuge
3. Shaker (rotatary)
4. Light and temperature controlled culture room
5. Magnetic stirrer
6. Balance
7. Stereoscopic microscope
8. Hemocytometer
9. Florescent microscope
10. Forceps, scalpel scissors etc.
11. Sterilized petri dishes of different sizes.
12. Sterilized pasteur pipettes
13. Sterilized centrifuge tubes
14. Sterilized beakers
15. Sterilized seive 64 μm.
16. Sterilized millipure filter and alongwith holders.
17. Parafilm
18. Sterilized small sized bottles (100 ml)

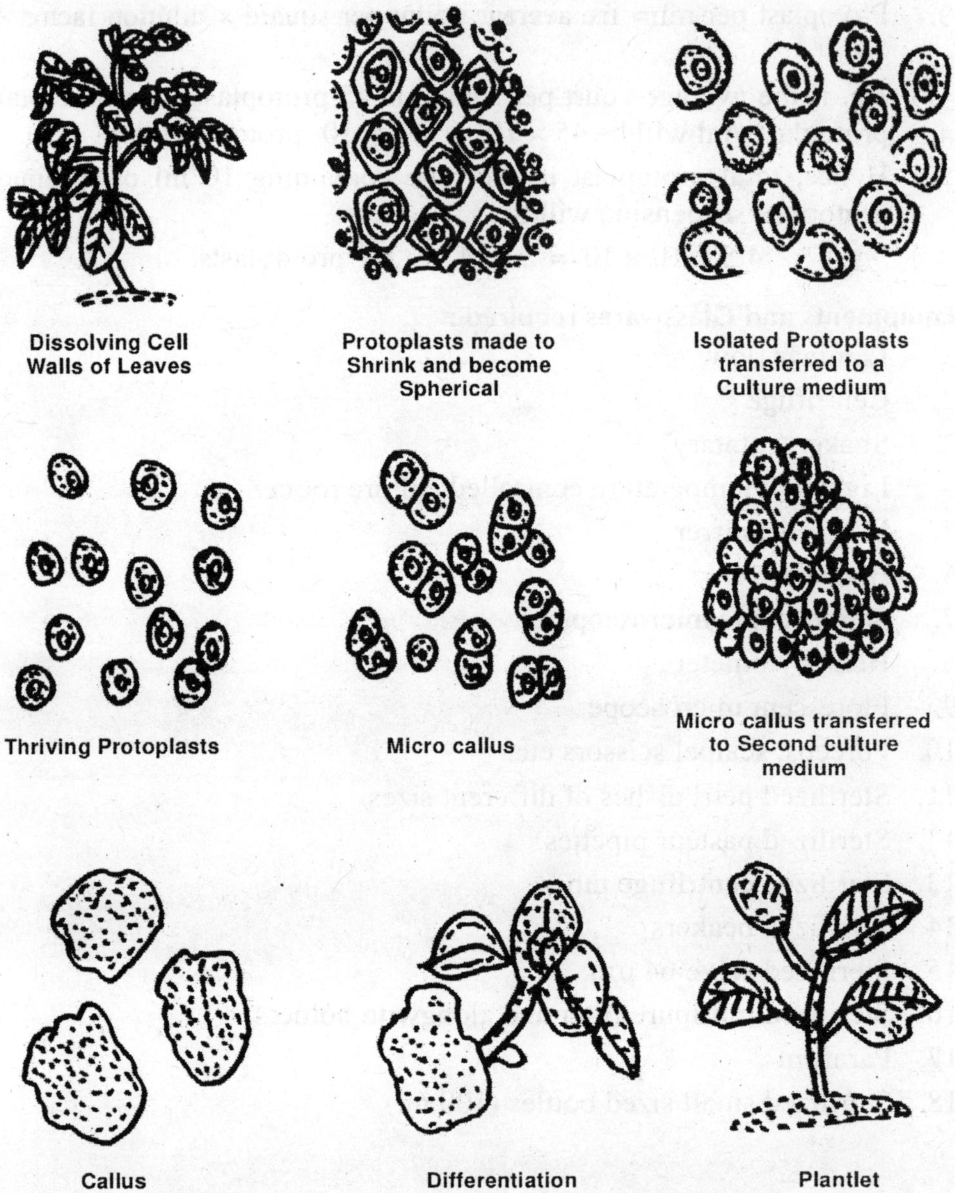

Fig. 16.1 Protoplast isolation and Plant regeneration

Protocols

1. Hypocotyl and leaves of *in vitro* raised aseptic seedling are used as explants to isolate protoplasts. (*See* protocals in Ch. 3)
2. Cut the leaves into 1mm wide strips, or small segments of hypocotyl. Place them in CPW salt solution with 9% w/v manitol (CPW 9M medium)

3. Transfer the plant tissues to a filter sterilized (0.22 μm) sterile enzyme mixture consisted of cellulase 2% + pectolyase 0.05% w/v, disolves in CPW 9M medium (pH 5.8).
4. Incubate at 25°C for 16 hours on a rotary shaker (30 cycles/min).
5. After incubation, the digested tissues are passed through a sterilized nylon sieve (64 μm pore size).
6. Give gentle washing in a small volume of CPW 9M medium using sterilized pasteur pipettes.
7. Fill the suspension, gently in sterilized centrifuge tube and centrifuge for 5-7 min at a speed of 1000 × g.
8. The sedimented protoplast are taken out by removing the supernatant.
9. Mix the sedimented protoplast with CPW 9M floated with CPW with 21% w/v sucrose. (CPW 21S medium) and centrifuge for 15 minutes at a speed of 1500 × g.
10. Protoplasts are collected in the middle of the centrifuge tube in the form of a green ring.
11. Then wash these protoplasts twice in CPW 9M medium followed by centrifugation each time for 7 minutes at a speed of 1000 × g.
12. Resuspend the protoplasts in K8P (use the same medium in which the protoplast are cultured) medium and centrifuge for 10 min at a speed of 100 × g.
13. Estimate the protoplast yield with the help of Haemocytometer.
14. Culture the protoplasts at a plating density of 2×10^5 protoplast in K8P (Kao, 1977) medium in small petri dishes containing 10ml of liquid medium. Seal the petri dishes with parafilm.
15. Incubate the protoplasts at 25°C under 16/8 for photoperiod.
16. Add 2-3 drops of fresh medium at an interval of 7 days for lowering down the osmotic pressure. Estimate the plating efficiency on 25th day of culture.
17. After 6-8 weeks of culture, clearly visible colonies/microcalli should be transferred on agar/agarose solidified MSP_1 (MS + 2.0 mg/l NAA + 0.5 mg/l BAP) and MSD_4 (MS + 0.05 mg/l NAA + 0.5 mg/l BAP) media for intensive callus formation.
18. Transfer the bigger calli on R (MS + 1.0 mg/l BAP + 0.1 mg/l GA3 + 500 mg/l as casein hydrolysate) medium for embryogenesis/organogenesis and subsequent plant regeneration.

Note: (i) Entire procedure is carried out strictly under aseptic conditions.

(ii) Use double distilled water for the preparation of media and other solutions.

(iii) Enzyme mixture required to isolate protoplast for some of the crop species are given in Table 1.

Table 1. Enzyme mixture and conditions for protoplast isolation

Species	Tissue	Enzyme mixture	Remarks
Arabidopsis thaliana	Leaves	1% (w/v) Cellulase R10, 0.25% (w/v) Macerozyme R10, 8mM $CaCl_2$, 0.4 M mannitol, pH 5.5	Cut leaves into 1mm^2 pieces, plasmolyse for 60 min in 0.5 M mannitol, digest for 14-18 h, 25°C, dark 16 h.
Hordeum vulgare	Embryogenic cell suspension	2% (w/v) cellulase Onozuka RS, 0.05% (w/v) pectolyase Y23, 976 mg/litre MES, CPW13M, pH 5.6	16h digestion
Brassica napus	Hypocotyl, stem, and petiole	1% (w/v) cellulase, 0.1% (w/v) Macerozyme R10, in K3 solution containing 136.9 g/litre sucrose, pH 5.6	Cut small pieces & plasmolyse for 1hr. digest for 14-16 hrs. at 25°C dark.
Nicotiana tabaccum	Leaf	4% meicelase+4% mecrozyme, pH 5.8, 13%, Mannitol	cut small pieces of 1 mm^2
	cell suspension	2% Driselase + 2% cellulase. pH 5.8	Digest for 14-16 hrs. at 27°C in dark.

Species	Tissue	Enzyme mixture	Remarks
Dimorphoteca aurantiaca	Leaves, seedling cotyledons	2% (w/v) Meicelase, 2% (W/V) Rhozyme HP 150, 0.03% (w/v) Macerozyme R10, CPW8M, antibiotics, pH 5.8	Plasmolyse for 1h, 5-6 h static incubation for leaves, 10-12 h for cotyledons
Dimorphotheca aurantiaca	Ray florets, callus, stem	0.8% (w/v) cellulase R10, 0.4% (w/v) Driselase,0.8% (w/v) Rhozyme HP 150, 0.4% (w/v) Macerozyme R10, CPW13M, antibiotics, pH 5.8	Plasmolyse for 1h, 10h static incubation for ray florets, 4-5 h shaki-ing followed by 8-10 h static incubation for stems and cotyledons
Glycine max	Hypocotyls from 5 to 7 days old seedlings	As for Dimorphotheca leaves, but dissolved in CPW13M solution	Plamolyse 30 min., incubates with enzyme for 16 h with shaking (10 r.p.m.) in the light 2.5 uE .m^{-2}.s^{-1})
Glycine canescens	Cotyledons from 10 days old seedlings	1% (w/v) Rhozyme HP 150, 2% (w/v) Cellulase R10, 0.1% (W/v) pectolyase Y23, Macerozyme R10, CPW13M, antibiotics, pH 5.8	Plasmolyse for 20 min, incubate for 16 h in the dark
Rudbeckia hirta	Leaves cotyledons, ray florets	0.8% (w/v) cellulase R10, 0.4% (w/v) Driselase, 0.8% (w/v) Rhozyme HP 150, 0.2% (w/v) Macerozyme R10, (CPW8M, antibiotics, (CPW13M, for ray florets), pH 5.8.	Plasmolyse for 1h, then 3-4 h enzyme incubation with Shaking, followed by 8 to 10h stationary incubation in the dark (23°C)

2. PROTOPLAST ISOLATION, CULTURE AND PLANT REGENERATION USING CELL SUSPENSION CULTURE OF RICE *(Oryza sativa)*

(Abdullah *et. al.* 1986; Fugimura *et.al.* 1985)

A. Initiation of cell suspension culture (*See* protocol in chapter 5)

Equipments and reagents

1. Exponentially growing cell suspension cultures three to seven days after sub-culture
2. Autocalved sieves constructed from stainless steel, nylon or polyester fabric with pore sizes of 500, 64, 45 and 30μm
3. CPW salts solution with 130 g/litre manitol (CPW13M)
4. Enzyme mixture : 1%W/V cellulase, 0.1% W/V pectolyase Y23, pH 5.8
5. 14 cm petri dishes
6. Sealing film e.g. Nescofilm or parafilm M
7. Sterile pasteur pipettes and volumetric pipettes
8. Orbital platform shaker (30 rpm 2 cm throw) with temperature control (27°C.)

B. Protocols for isolation and culture of protoplasts

1. Filter the cell suspension through a siceve of 500 μm pore size into a preweighted 14cm petri dish. Remove the medium with a sterile 10 ml pipette to leave the cells in the dish.
2. Re-weight the petri dish and add the enzyme mixture (10ml/g fresh weight of cells).
3. Seal the petri dish with Nesofilm or Para film M and incubate overnight (15h) in the dark at 27°C on an orbital shaker (30 r.p.m) followed by stationary incubation for 1h under the same conditions.
4. Sequentially filter the cell suspension/enzyme mixture through sieves of 64, 45 and 30μm mesh to remove undigested cells.
5. Collect the protoplasts in the filtrate, wash three times by centrifugation (80g, 10 min), and resuspend in CPW 13M solution.
6. Resuspend the protoplasts in KPR medium (K8P medium supplemented with an extra 300 μg/litre 2,4-D) at a density of 5.0×10^5/ml in a centrifuge tube. Heat shock by incubating in a water-bath at 45°C for 5 min followed by 30 sec in ice.

7. Wash the protoplasts twice by centrifugation at 80g, followed by resuspension in fresh KPR medium.
8. Resuspend the protoplasts at a density of 3.5×10^5/ml in KPR medium made semi-solid with 12g/litre Seal Plaque agarose (mix equal volumes of a 24g litre agarose solution with double strength K8P medium)
9. Culture 2 ml of protoplast suspension in 3.5 cm petri dishes, sealed with Nescofilm. in the dark at 27°C.
10. After 14 days, divide the agarose layers from each dish into four portions and transfer each portion to a separate 5cm petri dish containing 3 ml liquid KPR medium. Incubate in the dark at 27°C until cell colonies 0.5-10 mm in diameter have developed.
11. Transfer the protoplast-derived colones to 5×5 squares well plastic petri dishes (four colonies per well). The wells should each contain 2 ml MSKN medium (MS medium supplemented with 2.0 mg/litre NAA, 500 µg/litre zeatin, and 30g/litre sucrose, solidified by the addition of 12g/litre agarose). Incubate at 27°C in the dark.
12. After 7-14 days, transfer shoots with roots and coleoptiles to MS medium supplemental with 30g/litre sucrose and 4g/litre agarose. Incubate at 25°C in the light (2ug. $m^{-2}.s^{-1}$.16h photoperiod, daylight fluorescent tubes.)
13. Transfer rooted plants to 8 cm pots containing soilrite mixture or perlite and grow to maturity in a glass house (natural daylight supplement'd with 30 uE.$m^{-2}.s^{-1}$ of daylight fluorescent illumination, day and night temperature maxima of 30°C and 18°C respectively).

3. PROTOPLAST ISOLATION FROM Brassica juncea

(a) Media and solutions

(i) Murashige and Skoog (MS) basal medium

(ii) K3 (Kao's) medium with 0.4% agarose

(b) Preplasmolysis solution

MS major salts (half strength)

MS minor salts (half strength)

$CaCl_2.2H_2O$	0.5%
Mannitol	7.2%

pH 5.8

(Sterilization by autoclaving)

(c) Enzyme solution

MS major salts (Full strength)	
MS minor salts (Full strength)	
Mannitol	7.2%
Cellulase R-10 (Onozuka)	1%
Pectolyase Y-23	0.025% (0.25% Macerozyme R10 for leaf protoplasts)
pH 5.6	
(Filter sterilize)	

(d) 21S solution

Sucrose	21%
CaCl2	0.5%
MES	5 mM
pH 5.6	
(Sterilization by autoclaving)	

(e) Protoplast wash solution (W5)

NaCl	2.2 g
$CaCl_2$	4.6 g
KCl	0.100 g
Glucose	0.250 g
Mannitol	7.5 g
Make upto 250 ml (pH 5.6)	

(f) Protoplast dilution medium (Kz) major

	mg/1
NH_4NO_3	250
KNO_3	2500
$CaCl_2.2H_2O$	900
$MgSO_4.7H_2O$	250
NaH_2PO_4	150
$(NH_4)_2SO_4$	134
$CaHPO_4.H_2O$	50
Micro	
$CoCl_2.6H_2O$	0.025

$CuSO_4.5H_2O$	0.025
$Na_2MoO_4.2H_2O$	0.25
$FeSO_4.7H_2O$	27.8
Na_2 EDTA.$2H_2O$	36.3
KI	0.75
H_3BO_3	3.0
$MnSO_4.4H_2O$	13.2
$ZnSO_4.7H_2O$	2.0
Vitamins	
Nicotinic acid	1.0
Pyridoxin HCl	1.0
Inositol	100
Thiamine HCl	10.0
Sugars	
Xylcose	250
Sucrose	3.42%
Growth regulators	0.1 mg/l
2,4-D	1.0 mg/l
BAP	1.0 mg/l

pH 5.8

For plating use K_3 medium with 0.4% agarose

I. *In vitro* culture of Brassica juncea and B. oleracea seeds

(*All work should carried out in a laminar flow.*)

1. Surface sterilize *B. juncea* and *B.oleracea* seeds by treating with 0.1% $HgCl_2$, for 12-15 min.
2. Throughly wash treated seeds by rinsing (4-5 times) with sterile distilled water.
3. Transfer 8-10 seeds/tube containing half-strength MS-agar medium. (About 12 tubes will be needed for one isolation experiment).
4. Incubate in dark at 25° ± 2°C for 7 days

II. Isolation of protoplasts (Kartha *et.al.* 1974; Schenck and Hoffman, 1984)

1. Take 7 day old etiolated hypocotyls of *B. juncea* on a sterilized petridish.

2. with the help of a sharp scapel cut about 50 hypocotyls into small segments (5mm) and aseptically add 15ml of preplasmolysis solution containing 0.5% $CaCl_2$ and 7.2% mannitol (pH 5.8). Leave for Ih. Similarly repeat for leaf tissue.
3. Remove pre-plasmolysis solution with a sterile pasteur pipette and add 10-15 ml of enzyme mixture. Seal the dish with parafilm.
4. Incubate in dark at 25°C for 16-18 h.
5. Release the protoplasts by gently drawing through pasteur pipette (8-10 times).
6. Pass the enzyme mixture along with protoplasts and cell debris through a stainless steel mesh (250 µm).
7. Collect the filtrate and put in a 15 ml centrifuge tube and centrifuge at 300g for 5 min. (Use 100×g for leaf protoplasts)
8. Take and discard supernatant with the help of a pasteur pipete.
9. Suspend the pellet in 21S sucrose solution and spin for 10 min. at 300 g.
10. Collect protoplasts that float at the top with a pasteur pipette.
11. Dilute to about 1:10 with protoplast wash solution. Centrifuge at 300g (5 min).
12. Discard the supernatant and repeat washing once again.
13. After discarding supernatant, add 1 ml of protoplast culture medium (Kao's medium) to the pellet and suspend the protoplasts.
14. Draw a small aliquot of protoplast suspension and count the density of protoplasts with the help of a haemocytometer.
15. Adjust the density of protoplasts to 10^5/ml by adding protoplast culture medium.
16. Dispense 2 ml of protoplst suspension per 30mm petri dish. Seal and incubate in dark at 25°C (3-4 days).
17. Transfer to light (16h), 25°C after 4th day.
18. On 7th, 10th and 13th day add 0.5 ml of protoplasts dilution medium (K_3) to each dish. Incubate in light at 25°C. (After 1-2 days cell wall formation and cell divisions can be observed. Small cell clusters are visible by 7 days.)
19. After 19 days plate cells on agarose medium (K_3).

4. ISOLATION OF LEAF MESOPHYLL PROTOPLAST OF *B. oleracea* (Yamashita and Shimamato, 1989)

1. Excise leaves from asepticaly grown seedling of *B. oleracea* and peel off the epidermis, cut into small pieces and put into small petri dishes containing preplasmolytic solution & keep for about 1 h.
2. Take out the preplasmolytic solution with sterile pasteur pipette and add 10-15 ml enzyme mixture cellulose Onozuka R10 0.5% + pectolyase Y23 0.004% dissolved in 0.6M mannitol.
3. For isolation and culture of protoplast, follow step 5-19 of *B. jancea.*

5. PROTOPLAST ISOLATION AND CULTURE FROM TOBACCO LEAVES (*Nicotiana tabaccum*) (Banks and Evans, 1976; Bourgin *et.al.* 1979 and Nagata & Takebe, 1970)

1. Excised leaves from field grown seeding are rinsed in a soap/detergent solution followed by 3-4 washing and then surface sterilized in a 2% solution of sodium hyprochlorite/0.1% $HgCl_2$ for 20 min followed by 6-7 washing by sterilized distilled water. All subsequent operation are performed under aseptic conditions.
2. The lower epidermis of the sterilized leaves is peeled off using sterilized forceps and plastic gloves. Peeled leaves are cut into pieces of about 4 cm^2 using a scalpel/sterilized razor blade.
3. Put the pieces of leaves in small sterilized petridish containing preplasmolytic solution for about 1 hr.
4. Remove the preplasmolytic solution with sterile pasteur pipette and add 10-15 ml of filter sterilized enzyme mixture (Meicelase 40% +Mecro zyme 4%, dissolved in 0.6M manitol, pH 5-8) Sealed the petri dish with parafilm.
5. Incubate for 16-18 hours at 25 ± 2°C in the dark.
6. Watch the preisolated protoplasts under light microscape and pass the protoplast enzyme solution through 64μ sterilized sieve, under aseptic conditions to separate protoplasts from undigested cells.
7. Wash enzyme from protoplasts by centrifuging for 4min at 40 × g in a sterile centrifuge tube.
8. Decant and discard supernatant with sterile pasteur pipette. Gentaly resuspend the pelleted protoplast in 4-0ml of Kao and Michayluk (1975). K8P medium and repeat centrifugation and decantation steps.

9. Resuspand washed protoplasts in K8P medium (2.0 ml medium is added to 0.5 ml protoplasts pellets)
10. Place resuspend protoplasts in small droplets into 60 × 15 mm sterile petri dish. Seal the plates with double layer of parafilm to avoid contamination and dihydration of protoplasts. Incubate protoplasts under defuse light in plastic boxes.
11. First division of protoplasts should occur within 2-5 days after plating.
12. Add fresh medium K8P to droplets every 1-2 weeks untill rapidly proliferating callus is visible.
13. Shoot regeneration from callus is accomplished by adding MS medium supplemented with 5.0 μM BAP to liquid protoplasts derived suspension at 2 weeks intervals. Shoot should regenerate within 6 weeks intervals.
14. Regenerated shoots are rooted by transferring to half strength MS agar medium.
15. When an exclusive root system has formed, the regenerated shoots are transferred to peat pellets under high humidity for further growth.
16. Gradually decrease the humidity and increase the light intensity before transferring regenerated plants to green house conditions.

REFERENCES

1. Abdulah, R. Cocking, E.C. and Thompson, J.A. 1986. Biotechnology, 4 : 1087-1090.
2. Banks, M.S. and Evans, P.K. 1976. Plant Sci. Lett., 7 : 409-416.
3. Bourgin, J.P, Chupeau, Y. and Missonier, C. 1979. Physiol. Plant., 45 : 288-292
4. Fujimara, T., Sakurai, M., Akagi, H., Negishi, T. and Hirose, A. 1985. Plant Tissue Cult. Lett. 2(2) : 74-75.
5. Kartha, K.K. Michayluk, M.R. Kao, K.N.; Gamborg, D.L. and Constabel, F. 1974. Plant Sci. Lett., 3 : 265-271.
6. Kao, K.N. 1977. Mol. Gen. Genet., 150 : 225-230.
7. Punia, M.S. 1990. Ph.D. Thesis. Inst, of Genetics, (BAS) Sofia (Bulgaria).
8. Punia, M.S., Bohorova, N.E. and Atanassov, A.I.1994, Biot & Bio. Eq., 8(2) : 23-26.
9. Punia, M.S., Bohorova, N.E. 1996. In : Resource Management in Agriculture (R.C. Dogra, R.K. Behl and A.L. Khurana, Eds.) CCS HAU & MMB, New Delhi; pp. 201-205.

10. Nagata, T. and Takebe,I. 1970. planta, 99 : 12-20.

11. Schenck, H.R. and Hoffman, P 1984. In : Cell Culture and Somatic Cell Genetic of Plants (Vasil, I Ed.) Academic Press INC, Vol. 1 356-369.

12. Yamashita, Y. and Shimamoto, K.1989. In : Biotechnology in Agriculture and Forestry - Plant Protoplast and Genetic Engineering-1 (Y.P.S. Bajaj, ed.) Springer Verlag, Berlin. Vol 8 : 193-205.

17

PRODUCTION OF SOMATIC HYBRIDS THROUGH PROTOPLAST FUSION

Under appropriate conditions protoplasts which are naked cells (cells without cell wall) from different plants can be induced to fuse in culture, combining their genetic information to create a new hybrid. The recombinant protoplast can then be induced to reform cell wall, proliferate, form callus, and regenerate (Fig. 17.1). This technique can be used to fuse protoplasts of the different species or genera, which leads to the creation of entirely new hybrids from species that cannot be crossed sexually.

Two methods are most commonly used for protoplasts fusion *viz.* (i) chemical fusion using polyethylene glycol (PEG), and (ii) Electro-fusion using electric pulse. However, the PEG mediated fusion of protoplast is widely used method for somatic hybridization. The protocols for the fusion of Brassica juncea hypocotyl derived etiolated protoplasts and leaf mesophyll derived green protoplasts of *B. oleracea* are as follows :-

1. PEG MEDIATED FUSION OF PROTOPLASTS

Polyethylene glycol is a chemical fusogen widely used in somatic fusion of both plant protoplasts and animal cells. Application of a highly concentrated solution of PEG draws out H_2O from protoplasts and brings together the protoplasts such that the membranes of protoplasts come into close contact. When PEG solution is diluted, protoplasts enlarge and the membranes of adjacent protoplast fuse at the point of contact. Thus somatic fusion is accomplished.

Solutions required

1. Fusion solution

PEG 8000	20% (w/v)

DMSO	10% (w/v)
Glucose	3.6% (w/v)
$CaCl_2.2H_2O$	0.17% (w/v)
KH_2PO_4	0.0095% (w/v)
Filter sterilize	(pH 5.8)

2. Wash solution (W.)
3. Protoplast culture medium (Kao's medium).

Protocols

1. Isolate protolasts using protocols described in the previous chapter and suspend in culture medium at a density of 10^6/ml.
2. Mix equal volume of both the hypocotyl and the leaf mesophyll protoplasts of *B. juncea* and *B. oleracea* in a centrifuge tube.
3. Place two drops (40-50µl each) of protoplast mixture on a 30 mm sterile disposable petri dish. Allow protoplasts to settle for 5 minutes.
4. Gently introduce from one side equal volume of fusion solution to each of the droplets. Leave undisturbed for 10 min.
5. Draw out with a pasteur pitette most of the solution from each of the droplets and discard. (Take care not to remove all solution, protoplasts will collapse).
6. Add 40µl of wash solution (W_5 medium) to the protoplast droplets. Allow to stand for 2 min.
7. Repeat step 6 two more times. Allow to stand for 10-15 min.
8. With the help of a pasteur pipette draw out most of the wash solution and discard.
9. Add 1.5 ml of protoplast culture medium to each dish. Seal with parafilm.
10. Incubate in dark 25°C for 3-days (Follow steps 17-19 of protoplast culture for further culture protocol of fusion products.)

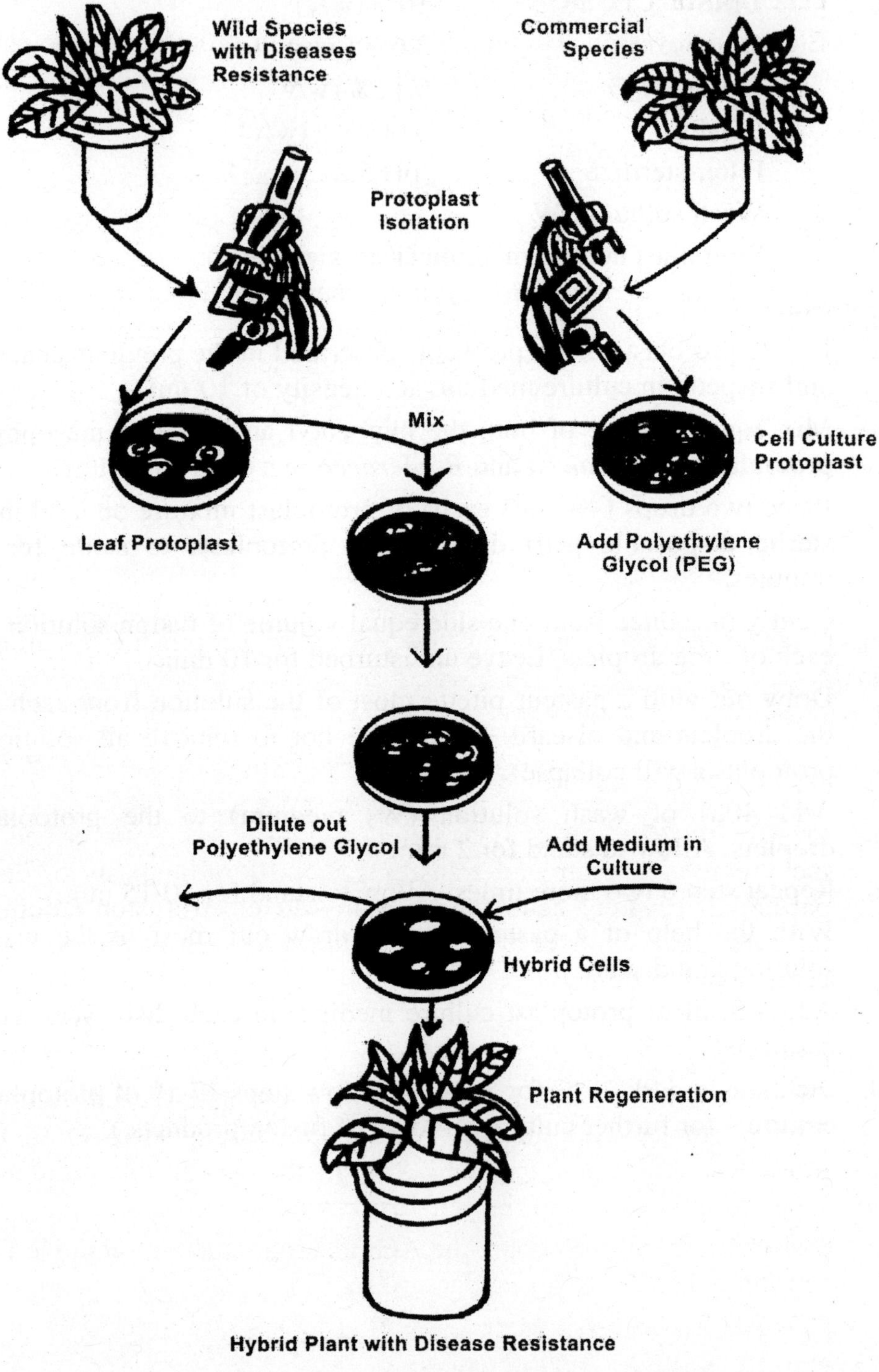

Fig. 17.1 Production of Somatic Hybrids

2. ELECTROCELL FUSION

Electro-cell fusion is a technique for fusion protoplasts through a physical process mediated by electric currents in contrast to the PEG mediated fusion where polyethylene glycol acts as a fusogen. In contrast to electroporation, which is a single step process mediated by DC pulse, electrofusion is a two step process mediated by AC frequency as well as a DC pulse. Generation of AC frequency of 1 MHz between electrodes in a near neutral medium leads to the formation of pearl chains in protoplast suspension facilitating close contact between adjacent cell membranes of protoplasts. After pearl chain formation has been achieved, membrane break-down between adjacent protoplasts would be achieved by a DC pulse as in electroporation. The equipment necessary for electrically induced fusion is commercially available.

Protocols

This protocol describes the use of the electro fusion apparatus which employs a parallel plate electrode system. Protoplasts can be observed during fusion using an inverted microscope inside a laminar air flow cabinet.

1. Disconnect the electrode from the apparatus and sterilize by immersion in 70% (v/v) ethanol solution for 10 min. Allow the electrode to dry inside the air stream of the flow cabinet.
2. Prepare suspensions of both parental protoplasts as described in the previous chapter and transfer 1.5×10^6 protoplasts of each partner to separate centrifuge tubes. Add electrofusion solution, 0.5 mM $CaCl_2$ with 11% (W/V) mannitol to each of the tube and centrifuge (15 min, 100g).
3. Remove the supernatant from each tube and resuspend each protoplast pellet in 15ml of electrofusion solution.
4. Dispense 0.75 ml of each protoplast preparation into each well of a 5 × 5 square well plastic Petri dish, leaving two wells for protoplast self-fusion, Pipette 1.5 ml of the appropriate protoplast suspension into each of the self-fusion wells.
5. Switch on the oscilloscope, the AC function generator, and the DC supply, Select the required parameters :
   * AC frequency, 1 MHz
   * DC voltage, 1000 (V/cm)
   * number of pulses, 4

* Pluses width, 1.5 msec
* inter-pulse separation, 1 sec.

6. While observing the protoplasts, increase the AC amplitude and monitor pearl chain formation. Occasionally, spinning of the protoplasts is observed, in which case the AC frequency should be altered.
7. Increase the AC amplitude to maximum to induce close membrane contact between adjacent protoplasts and immediately activate the DC pulse sequence.
8. Reduce the AC filed to the alignment amplitude, followed by a gradual reduction to zero.
9. Transfer the electrode to the next well and repeat the sequence, altering the DC parameters as necessary. If excessive bursting to the protoplasts is observed, the pulse width and DC voltage should be reduced. Conversely, these parameters should be increased if the fusion frequency was previously seen to be low (less then 5%). Frequently, large multiple fusions occur. This can be avoided by reducing the time of alignment in the AC field and by reducing the protoplast density.
10. Following fusion treatment, add 1.0 ml of culture medium to each well and leave the plate undisturbed for 1 h to allow the heterokaryons to stabilize. Subsequently, remove 1.0 ml of medium from each well and add 1.0 ml of fresh culture medium. After a further 15 min incubation at 22°C, transfer the protoplasts to a centrifuge tube, mixing the contents of each well thoroughly to resuspend the protoplasts which have settled onto the base of the dish. After centrifugation and resuspension in medium, culture the protoplasts as described in the previous Chapter.

Note: 1. Reduce the osmotic pressure of the medium as low as possible without causing protoplasts lysis, since this allow an increase in the protoplast diameter leading to greater voltage across the protoplast membrane.

2. Application of multiple pulse of shorter duration is frequently preferable to a single long pulse.

3. SELECTION OF HETEROKARYONS AND SOMATIC HYBRID TISSUES

The most difficult aspect of the somatic hybridization procedure is the selection of heterokaryons, heterokaryon-derived cells or tissues, or hybrid plants. Several approaches have been devised. Some workers have cultured all protoplasts subjected to the fusion treatment and screened the regenerated plants for their somatic hybrid or cybrid characteristics. The disadvantage of this procedure is that it is labour intensive and considerable growth room/glasshouse space may be required to house regenerated plants. Occasionally, heterokaryon-derived cells may exhibit heterosis and outgrow the cell colonies derived from parental protoplasts or homokaryons. Such faster growing colonies can be isolated mechanically and eventually transferred to plant regeneration medium. Other procedures which have been used successfully include manual selection of heterokaryons and high technology flow cytometry.

Manual Selection

When the two parental protoplast populations can be easily identified, it is possible to pick out individual heterkaryons with micropipettes. This arrangement works well in the fusion of chlorophyll -containing leaf protoplasts with colourless protoplasts isolated from cell suspensions. A simple micromanipulation device is available. A major advance in this field has been the development of a dual fluorescent labelling system for heterokaryons. Protoplast labelled green by treatment with FDA (1-20 mg/l) are fused with ptrotoplasts emitting a red fluorescence, either from chlorophyll auto fluorescence or from exogenously applied rhodamine isothiocyanate (10-20 mg/litre;). Although micromanipulation of individual heterokaryons is both time-consuming and tedious, it is technically relatively straightforward and is a reliable method for somatic hybrid production.

Complementation

Procedures have been developed which inhibit growth of homokaryons and ideally both, but certainly one, of the parental protoplast populations. Mutants exhibiting hormone autotrophy, auxotrophy, resistance to antibiotics, amino acid analogues, and fungal toxins have been used in various combinations, or with partners which have been inactivated by X-ray or uv-irradiation, or by treatment with iodoacetamide. Albino mutants have also been employed successfully and hybrid tissues identified by their chlorophyll production on exposure of cultures to light.

Flow Cytometry

Cytometry can also be used to physically isolate heterokaryons. This technique uses the same method of identification of the heterokaryons, i.e. dual fluorescent labelling. Inside the flow cytometer, the protoplasts are passed in a flow of liquid between the light source and the fluorescence detectors. The stream is then dispersed into droplets and the computer electrostatically deflects the droplets containing heterokaryons into a sterile tube. The procedure is fully automatic and rapid; upto -2000 protoplasts per second can be analysed. However, the instrument is extremely costly and for some protoplast systems, only about 10% of the heterkaryons are retained. This technique is normally used when it is possible to produce and fuse very large numbers of protoplasts (5×10^7) of each parental population.

4. CHARACTERIZATION OF SOMATIC HYBRID PLANTS

Somatic hybrid plants have been characterized morphologically, cytologically, and biochemically. Routine biochemical characterization includes analysis of isozymes and fraction 1 protein, the resistance of plants to viral infection and antibiotics, or sensitivity to herbicides and fungal toxins. Genetic analysis can be performed, provided hybrid plants are fertile. More recently, the molecular techniques of RFLP and RAPD analyses have been employed while flow cytometry provides a method for rapid analysis of the nuclear DNA content for ploidy determinations.

REFERENCES

1. Aviv, D and Galun, E.1988, Mol. Gen.Genet, 215:128-133.
2. Bonnema, A.B., Melzer, J.M. and O'Connel, M.A. 1991, Thoer. Appl. Genet, 81:339-348.
3. Menezel, L., Nagg, F. Kiss, Zs. R. and Maliga, P.1981. Theor. Appl. Genet; 59:191-195.
4. Perl A., Aviv, D. and Galun, E. 1991 Mol. Gen. Genet, 225:11-16.
5. Punia, M.S. 1990. Ph. D. Thesis. Inst. of Genetics (BAS), Sofia (Bulgaria)
6. Sundberg, E., Lagercrantz, U. and Glimelius, K. 1991, Plant Sci. 78:89-98.
7. Tample. M., Makaroff, C.A., Mutschler, M.A. and Earle, E.D. 1992. Curr. Genet, 22:243-249.
8. Water, T.W. Mutscher, M.A. and Earle, E.D. 1992. Plant cell reports, 10:624-628.

18

ELECTROPORATION OF PROTOPLASTS

Electroporation is a physical reversible process leading to the formation of transient pores in the cell membrance of the protoplasts which would allow the transfer of nucleic acids and other macromolecules inside. In general, genetic modification in prokaryotes is referred to as transformation and in eukaryotes, the process is called transfection. Since it is based on electrical discharge, it is more appropriately called electro-transfection. This reversible process of creating pores in cell membrane is mediated by a DC pulse discharged by a capacitor in an electroporation. The protocols developed for tobacco mesophyll protoplasts are described in this chapter.

Protocols

1. Wash the purified tobacco mesophyll protoplasts with W5 solution.
2. Wash protoplasts once again in 0.6M mannitol solution at pH 7.0-7.5 and centrifuge at 100 × g for 3-5 minutes.
3. Suspend the pellet in 0.6M mannitol at 10^6 protoplasts per ml.
4. Dispense 250µl of protoplast suspension to the pre-sterilized electroporation cuvette with absolute alcohol for atleast 15 minutes.
5. Sterilize electroporation electrodes with absolute alcohol for atleast 15 minutes.
6. Add 5-10µg DNA (pBI 221 with-GUS)
7. *Electroporation settings*

Choose mode	500V
Resistance R	R9 (480 ohms)
Capacitance C	100µF
Cuvette gap	2mm
Field strength	0.5kv/cm
Pulse duration	32msec.

8. After electroporation, keep protoplasts on ice for 10 minutes before plating, on K8P medium, using different plating methods.

REFERENCES

1. Fromm, M, Taylor, L.P., and Walbot, V.1986. Nature, 319:719-793.
2. Paszkowski, J., Shillito, R.D., Save, M.W., Mandak, V., Hohn, T., Hohn, B. and Potrykus, I. 1984, EMBOJ; 3:2717-2722.
3. Shillito, R.D., Soul, M.W., Paszkowski, J. Muller, M. and Potrykus, I.1985. Bio/technology, 3:1099-1103.

19

GENETIC TRANSFORMATION OF PLANTS BY USING *AGROBACTERIUM TUMEFACIENS* VECTOR SYSTEM

One of the most facinating achievements in plant molecular biology is the possibility of changing the genotype and phenotype of a plant by introduction of a defined set of genes in its genome. Although there are a lot of different methods for plant transformation, the most widely used and effective one is the use of the unique natural gene transfer capabilities of the soil borne plant pathogenic bacterium *Agrobacterium tumefaciens*. Unlike other plant pathogens it transfers a defined part of its tumour inducing (Ti) plasmid into the genome of a plant cell. This part of the Ti plasmid is called T-DNA and it contains genes forcing the plant cell to divide and form a callus mass. Another set of transferred called opines by the transformed plant cells. Another region of the Ti plasmid, the virulence (vir) region encodes most of functions necessary for T-DNA transfer.

The polarity and position of 25 bp long bordering the T-DNA are essential for its transfer and integration into thc plant genome. These processes are not dependent on the expression of the T-DNA encoded genes, but are co-ordinated by the expression of genes in the virulence region. The use of *Agrobacterium* for genetic engineering in plants is based on the fact that any DNA sequence which is between the T-DNA borders can be transferred to the plant genome. Since the early experiments on transformation with genetically engineered *Agrobacterium* strains, efforts has been made to broaden the application possibilities of the system. The binary vector system is composed of two elements: one of them is a small, easy to manipulate plasmid that contains only the T-DNA borders and between them a selectable marker gene for Positive selection of the rare transformed plant cells, a multiple cloning site, and something other elements like lambda cos sites for recloning the transformed DNA. This binary vector can replicate both in *Esherichia coli* and in *Agrobacterium*. The second element of the system is a large Ti plasmid without functional T-DNA needed to provide the transfer. In the case of the binary system the two plasmids replicate in *Agrobacterium* independently. The "disarmed" (a wild type) Ti plasmid provides the virulence functions in trans. The intermediate vector system or cointegrated vector system uses manipulated Ti plasmid which can integrate cloned genes from a donor plasmid within their T-DNA region by homologous recombination.

The main procedures carried out for the introduction of a foreign gene into plants using a *Agrobacterium* binary vector system are :

1. Cloning of the gene of interest with the appropriate regulating sequences within the T-DNA border in a binary vector.
2. Mobilizing the binary vector into an *Agrobacterium* host strain by triparental mating. Verifying the structure and integrity of the plasmid in the transconjugants.
3. Inoculating the plant tissue with the verified *Agrobacterium* strain.
4. Selecting of putative transformants.
5. Analysis of the regenerants.

1. MOBILISATION OF A RECOMBINANT PLASMID FROM *E. COLI* INTO *A. TUMEFACIENS* VIA TRIPARENTAL MATING (CONJUGATION)

Media and antibiotics

<u>LB medium (per L)</u>

Yeast extract	5 g
Tryptone	10g
NaCl	10g

pH to 7.0-7.2

<u>ABM medium</u> (20 × AB salts + 20 × AB buffer)

<u>20 × AB salts</u> (per L)

NH_4Cl	20g
$MgSO_4.7H_2O$	6g
KCl	3g
$CaCl_2$	0.2g
$FeSO_4.7H_2O$	50mg

<u>20 × AB buffer (per L)</u>

K_2HPO_4	60g
$NaH_2PO_4.H_2O$	23g

<u>Mannitol solution (per L)</u>

Mannitol	10g

Autoclave the solutions separately and mix 50ml 20 × AB salts, 50 ml 20 × AB buffers and 900 ml mannitol solution to make 1L.

<u>MGL medium (per L)</u>

Tryptone	5g
NaCl	5g
Yeast extract	2.5g
Monosodium glutamate	1.2g
Mannitol	5g
K_2HPO_4	0.25g
$MgSO_4.7H_2O$	0.1g
Biotin	1 mg
pH 7.0	

PROCEDURE

1. Start three bacterial cultures in 3-5 ml LB medium with the appropriate antibiotics : donor *E. coli* strain carrying the binary vector, a helper strain HB 101 (pRK 2013) and an acceptor strain of *Agrobacterium*. Grow them overnight with shaking at 28-30°C for *Agrobacterium* and at 37°C for E.coli.
2. Wash the bacteria by dilution with LB medium, spin down for 10 min at 4000 rpm and resuspend the pellet in 3 ml antibiotic-free LB medium. This step could be omitted if no antibiotics have been added to cultivation medium.
3. Plate 0.1 ml of a 1:1:1 mixture from the three cultures on a MGL plate and incubate overnight at 27-29°C. Plate also control combinations of acceptor-donor, acceptor-helper and donor helper strains.
4. Collect single colonies from conjugation mixtures and plate them on solid ABM medium containing the appropriate antibiotics necessary for the growth of acceptor *Agrobacterium* and the binary vector. Transconjugants will appear in 2-3 days. The control combinations should not grow.
5. Purify single cell colonies by streaking on the same medium.
6. Verify the purity of the strain by *Agrobacterium* plasmid quick screen.

2. *AGROBACTERIUM* PLASMID QUICK SCREEN

Solutions

Solution 1

Lysozyme	4mg/ml
Glucose	50 mM
EDTA	10 mM
Tris. Cl pH 8.0	25 mM

Add the lysozyme just before use.

Solution 2

SDS	1%
NaOH	0.2 N

Freshly prepared

Solution 3

Sodium acetate, pH 4.8	3M

TE buffer

EDTA	1mM
Tris. Cl, pH 8.0	10 mM

PROCEDURE

1. Grow *Agrobacterium* cells overnight in 2-3ml LB medium with appropriate antibiotics at 28°C with vigorous shaking.
2. Transfer 1 ml to a 1.5 ml Eppendorf tube. Spin the cells down in an Eppendorf centrifuge for 30 sec.
3. Discard the supernatant and resuspend the cells in 0.1ml of ice-cold Sol. 1. Resuspend very carefully, cell clumps should not remain. Leave for 10 min at room temperature.
4. Add 0.2ml of Sol. 2 and shake well to mix. Incubate for 10 min at room temperature.
5. Add 30 µl Phenol equilibrated with Sol. 2 Vortex gently for a few seconds. The mixture will turn very viscous.
6. Add 150µl Sol. 3, Shake briefly to mix and incubate for 15-20 min at -20°C.
7. Spin the tube for 3 min quickly pour the supernatant into a new one.
8. Fill the tube with ice-cold ethanol. Invert several times to mix and store it at -80°C for 15 min.

9. Spin again for 3 min. Discard the supernatant and inverted on a paper towel to dry away the remaining solution.
10. Add 1ml ice-cold 70% ethanol, vortex and spin down for 2-3 min. Dry again as described above.
11. Resuspend the DNA pellet in 50µl TE buffer. The plasmid DNA is ready for analysis by restriction endonuclease digestion and fragment separation. Usually some chromosomal and helper plasmid DNA's are present but they rarely interfere with the analysis. The samples should be stored at -20°C

2. MAINTENANCE AND CULTIVATION OF BACTERIAL STRAINS.

Solutions

YEB medium (per L) :

bacto beef extract	5g
yeast extract	1g
peptone	5g
sucrose	5g
$MgSO_4$	0.492g
bacto agar	15g
pH 7.2	

Agarose for *Agrobacterium* (per L) :

Bactortripton	10g
yeast extract	10g
glucose	2g
Difco agar	2g

Agarose for *E.coli* (per L) :

nutrient broth	10g
NaCl	5g
Difco agar	6g

PROCEDURE

Isolation of single colonies

1. Sterilise a platinum transfer loop by flaming until it glows light red. Allow it to cool in the air or by stabbing it into a sterile agar medium. Use the cooled loop to pick up bacteria.
2. Touch the loop to a single well-isolated bacterial colony growing on the surface of YEB solid medium.

3. Streak the bacterial colony adhering to the loop onto a segment of a plate containing YEB agar medium. Sterilize the loop by flaming and cool it by stabbing into a region of the agar medium that is free of bacterial cells. Pass the loop once across one end of the primary streak and spread the bacteria that adhere to the loop into a fresh on the agar medium. Sterilize and cool again the loop and streak from one end of the secondary streak. Repeat step 3 twice.
4. Replace the lid of the plate. Label the bottom of the plate and incubate it in an inverted position at an appropriate temperature for 16-24 hrs. (*A. tumefaciens* 28°C, *E.coli* 37°C). Well separated colonies should be visible in the area of the final streak.

Short-term storage of bacterial strains

Colonies of most strains of bacteria can be maintained for periods of a few weeks on the surface of agar media if the plates are tightly wrapped in parafilm and stored inverted at 4°C.

Long-term storage of bacterial strains

Most strains of bacteria can be maintained for a year or two in stab cultures. Such cultures are usually prepared in small screw-capped bottles containing 2-3ml of agar medium or glycerol. Bacteria can be stored for many years in media containing 50% glycerol at low temperature without significant loss of viability. Liquid overnight culture of the bacterial strain (0.5ml) is mixed with the same quantity of glycerol in screw-capped vials and stored at -20°C to -70°C.

3. PREPARATION OF BACTERIAL STRAINS FOR TRANSFORMATION

A well developed colony from bacterial strain (A. tumefaciens) is suspended in 10ml liquid YEB medium supplemented with the antibiotic necessary for its growth in proper concentration (Table 1).

The overnight bacterial suspension is poured out in tubes and centrifugation at 5000 rpm for 5 min or at 2000 rpm for 20 minutes. Washing of the bacteria is carried out by re-suspending the pellet in liquid plant medium and centrifugation at 5000 rpm for 5 min or at 2000 rpm for 20 minutes.

The rate of growth of a bacterial culture is most conveniently monitored by withdrawing aliquots at various times and reading the optical density (OD) at a wavelength of 580 nm.

As a rough guide:

$0.8\ OD_{580}$ - 5×10^8 - 1×10^9 cells/ml (*A. tumefaciens*)
$1\ OD_{600}$ - 8×10^8 cells/ml (*E.coli*)

Aeration of cultures is best achieved by attaching a flask or tube to a rotary shaking platform. The volume of the vessel should be at least four times larger than the volume of medium it contains so as to allow vigorous shaking (220 rpm).

Table 1. Antibiotic stock solution for *Agrobacterium* strains.

Name	Stock mg/ml	Final concentration µg.ml
Rifampicin (Rif)	10	100
Carbeniciline (Cb)	50	100
Kanamycine (Km)	50	50
Tetracycline (Tc)	10	10
Streptomycine (Sc)	50	100

Note: All antibiotics dissolve in H_2O except Rif (Methanol, DMSO) and Cb (Ethanol). All antibiotic solutions are filter sterilized and stored at -20°C.

4. LEAF DISK TRANSFORMATION

Inspite of the successful gene transfer into several model species, the transformation of important crops is still a difficult task. Numerous factors are affecting the efficiency of the transformation. Among them the most crucial ones are:

1. Matching of the bacterial and plant genotypes;
2. An appropriate explant with available regeneration system;
3. Choice of an inoculation method;
4. Proper use of the selection pressure;

One of the most widely used transformation method is cocultivation of leaf explants (discs) with an *Agrobacterium* strain carrying a binary vector. For this method, usually 2-3 weeks old *in vitro* grown plants are used. It is possible to use greenhouse grown plant material which has been sterilized before starting the procedure of cocultivation.

PROCEDURE

1. The sterile leaves are cut in slices or discs with diameter 5-6mm in water (culture media or on wet sterile filter paper).

2. Leaf slices are immersed in bacterial suspension for serval minutes and without blotting the excess of bacteria are plated on solid regeneration medium without antibiotics.
3. Co-cultivation takes place at dim light at 25-28°C for 24-28 hours until the growth of bacteria is visible around co-cultivated explants.
4. After cocultivation, leaves are washed in sterile water blotted in filter paper to dry and plated on regeneration medium supplemented with 100µg/ml kanamycine and 500µg/ml calforan. Plates are co-cultivated in light at 25°C.
5. In 25-30 days green buds and then shoots are formed on cut leaf surface.
6. Shoots are cut and transplanted on phytohormone free medium with 100µg/ml kanamycine and 300 µg/ml calforan for rooting. Selective media should be changed every 3 weeks.

Note: Manipulation of sterile plant material, cocultivation procedures, washing are carried out in a laminar air flow.

REFERENCES

1. Anon. 1995. Manual of ICGEB practical course on Gene Transfer for Plant Breeding Improvement. Inst. of Genetic Engineering, Kostinbrod (Bulgaria)
2. An, G. Ebert, P., Mitra, A. and Ha, S. 1988. Plant Mol. Biol. Manual A3 : 1-19.
3. Brisson, M, Paszkowski, J, Penswick, J.R. Gronenborn, J. Potrykus, I., and Hoher, T. 1984. Nature, 310:511.
4. Davey, M.R. Cocking, E.C. Freeman, J. Perce, M, and Tudor, I, 1980. Plant Sci. Lett. 18 : 307
5. Ditta, G., Stanfield, S., Carbin, D. and Helinski, D. 1980. PNAS, 77 : 7347-7351.
6. Hoekema, A, Hirsch, P.R. Hooykass, P.J.J. and Schipleroort, R, A 1983. Nature, 303 : 179.
7. Horchh, R, Fry, J. Nedermeyer, J. Rogers, S and Fraley, R. 1988. Plant Mol. Biol. Manual AS : 1-9.
8. Mcbride, K.E and Summerfelt, K.R. 1990. Plant Mol-. Biol. 14 : 269.

20

AGROBACTERIUM TUMEFACIENS MEDIATED TRANSFORMATION

The present protocol describes the methodology for transformation of tobacco by co-cultivation method using a disarmed *Agrobacterium tumefaciens* (Ti plasmid) strain LBA 4404 containing a binary vector p35S GUSINT. This binary vector (Fig. 20.1) contains a kanamycin resistance gene (npt II) and a β-gulcuronidase gene (GUS) as a selectable and scorable marker, respectively.

1. CO-CULTIVATION METHOD OF GENETIC TRANSFORMATION

Requirements

Agrobacterium tumefaciens strain LBA 4404 (p35SGUSINT)

Axenically grown tobacco (*Nicotiana tabacum*) plants

Ethanol	
Kanamycin	(100 mg/ml)
Cefoxtaxime	(250 mg/ml)

Liquid medium

YEMB	
K_2HPO_4 solution	10 ml
Mannitol	10g
Yeast extract	1g
$MgSO_4$ (1M)	0.8 ml
K_2HPO_4 solution	
K_2HPO_4	5g
NaCl	1g
H_2O	100 ml

MS_0 (MS medium without phytohormones)

MS 104	1.0 mg/L BAP
MSO	0.1 mg/L NAA

Solid medium

YEMA	1.6 Agar/100 ml YEMB
MS104 (S)	0.7g Agar in 100 ml MS104
MS_0(S)	0.7g Agar in 100 ml MSO

Selection medium

100 mg/L kanamycin + 250 mg/L cefotaxime in MS104 (S)

Protocols

1. Grow the *Agrobacterium tumefaciens* in 5ml of YEM medium overnight at 28°C.
2. Tissues from aseptically grown sterile plants are most commonly employed. Tissues from non-aseptically grown plant must be surface sterilized before use.
3. Chop the tissue into small pieces (about 0.5 cm^2 sections for leaf tissue and about 0.5 cm long section for stems and hypocotyls.)
4. Place several pieces of plant tissue in a small petri dish (50 mm) Add 4 ml of MS 104 medium.
5. Add 50µl of an overnight grown culture of *Agrobacterium.*
6. Seal the petri plate with parafilm and co-cultivate bacteria with plant tissue for 2 days at 28°C in the dark.
7. Wash the plant tissue with MS_0 medium and bolt them dry with sterilized filter paper,
8. Incubate the tissue for a few days on a non-selective MS104 (S) medium containing 200 µg/ml cefotaxime.
9. After 5-7 days, transfer the tissue to selective medium MS 104 containing 100 µg/ml kanamycin and 200 µg/ml Cefotaxime.

Note : Manipulation of sterile plant material, co-cultivation procedures, washing are carried out in a Laminar flow box.

2. CHARACTERIZATION OF TRANSFORMANTS

The transformed nature of plant/plant tissue should be confirmed by performing a Southern hybridization analysis by using marker gene DNA as probe and by studying the expression of a reporter enzyme (e.g. β-glucuronidase).

A. Southern Hybridization Analysis

Protocols

1. Isolate high molecular weight DNA from transformed and untransformed plant tissues as described in the later chapters.
2. Digest the plant DNA with EcoRI and electrophorese the uncut and digested DNA on 0.7% agarose gel.
3. Carry out blotting of this gel on a nylon membrane.
4. Preparation of probe : plasmid carrying nptII gene. Isolate and purify this plasmid as described in the later chapters. For preparing probe go for nick translation of the whole plasmid using radiolabelled dCTP.
5. Carry out hybridization of the blot.
6. Expose the membrane to the film.
7. Develop the autoradiogram.

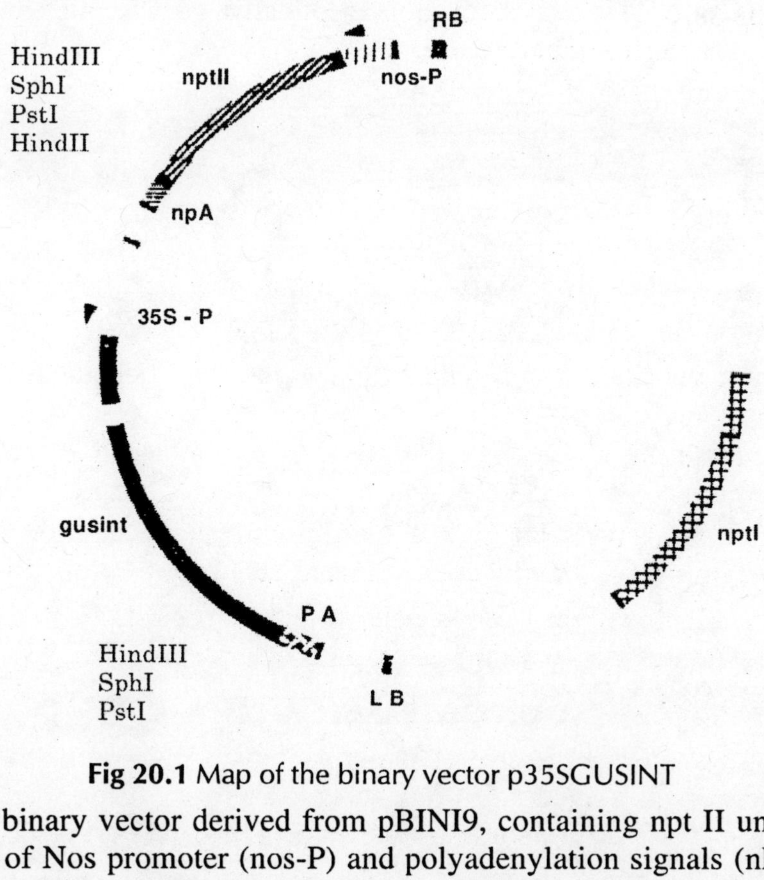

Fig 20.1 Map of the binary vector p35SGUSINT

A binary vector derived from pBINI9, containing npt II under the control of Nos promoter (nos-P) and polyadenylation signals (nPA)and an intron containing gus reporter gene (gusint) under the control of

CaMV35S promoter (35S-P) and polyadenulation signals (PA). It has the right (RB) and left border (LB) sequences of T-DNA from pTi37. The gene npt I, conferring kanamycin resistance can be use as a selection marker in bacteria.

REFERENCES

1. Anon. 1998. Training Manual of Recent Technique in Plant Biotechnology and Molecular Biology. NRC on plant Biotechnology, IARI, New Delhi.
2. Beran, M.H 1984. Nuleic Acids Res. 12 : 8711-8721.
3. Ditta, G. stanfield S., Corbin, D, and Helinski, D.R. 1980. Proc Natl Acad Sci. USA, 77 : 7347-7351.
4. Gallois, P. and Marinho, P., 1995. In : Methods in Molecular Biology : Plant Gene Transfer and Expression Protocols (Jens, H.ed.) Humana Press inc. Totowa NJ., Vol, 49 : 39-48.
5. Shaw, C.H. 1995. In Methods in Molecular Biology : Plant Gene transfer and expression protocols (Jones, H. ed) Humana Press Inc. Totowa N.J. Vol 49. 33-38.

21

ASSAYING MARKER GENES IN PLANTS - GUS (β-GLUCURONIDASE) GENE

The best substrate currently available for histochemical localisation of β-glucuronidase activity in transformed tissues and cells is 5-bromo-4-chloro-3-indolyl glucuronide (X-gluc). This substrate works very well, giving a blue precipitate at the site of enzyme activity. There are numerous variables that affect the quality of the histochemical localisation, including all aspects of tissue preparation and fixation as well as the reaction itself (1,2, 4). The product of reaction is not coloured. Instead the indoxyl derivative produced must undergo an oxidative dimerization to form the insoluble and highly coloured indigo dye (3).

GUS ASSAY-HISTOCHEMICAL

Solutions

1. 0.5M MES pH 5.6

   For 100ml: Dissolve 9.76g of MES in 80ml H_2O. Adjust pH to 5.6 with NaOH and make the volume to 100 ml. Store at room temperature.

2. Fixation solution (0.3% formaldehyde, 10mM MES pH 5.6, 0.3M mannitol)

   For 100 ml

Formaldehyde (40% sol. of BDH)	750μl
0.5M MES	2.0μl
Mannitol	5.46g

   Store in fume hood at room temperature for 3 month or until precipitation appears.

3. 50mM $NaPO_4$ pH 7.0 (Washing buffer).

   Add 57.7 ml of Na_2HPO_4 (1M stock)

42.3 ml of NaH_2PO_4 (1M stock)

100 ml distilled water to 200 ml

4. Histochemical staining with X-gluc.

   Dissolve 5mg X-gluc in about 100 µl dimethyl formamide Add 50mM $NaPO_4$, pH 7.0 to 5-10ml

   Make up in fresh when required. Store solid X-gluc dedicated at 20°C or -70°C 5.70% ethanol.

Protocols

1. Using a scalpel or a razor blade cut thin (0.51 mm) sections from plant leaves or stems.
2. Place these tissue sections in 200 µl of fixation solution and vacuum infiltrate.
3. Incubate at 30°C for 20 min.
4. Wash the tissue sections throughly (5-6 times) in washing buffer.
5. Incubate the sections in X-gluc-assay buffer (100 µl) at 37°C, overnight.
6. Rinse the tissue in 70% ethanol for 10 min.
7. Mount the tissue on a microscope slide and view under a microscope. Blue staining confirm transformation.

2. GUS ASSAY - SPECTROPHOTOMETERICAL

Solutions

GUS extraction buffer

50mM Na_2HPO_4, pH 7.0

10mM beta-mercaptoethanol

10mM EDTA

Stock solutions	Volumes (1L)
1M Na_2HPO_4, pH 7.0	50 ml
B-mercaptoethanol	0.7 ml
0.5M EDTA, pH 8.0	20 ml

Assay buffer (MUG) - Substrate

lmM MUG in extraction buffer. Dissolve 22mg 4-methylumbelliferone β-D-glucuronide in 50ml GUS extraction buffer.

Stop buffer

0.2M Na_2CO_3. Dissolve 21.2g Na_2CO_3 in 1L of distilled water.

Calibration standards.

1mM methylumelliferone. Dissolve in 10ml distilled water. Wrap the bottle in alumminium foil and store at 4°C.

Add 100 μl ImM stock solution to 900 μl stop buffer, mix well. 10μM MU: Add 100μl of this solution to 9.9ml stop buffer

100nM MU: Add 1.0ml of 1μM MU to 9.0ml stop buffer.

Protocols

1. Grind the tissue using a mortar and pestle in extraction buffer and freezing the tissue it in liquid nitrogen (use 2ml of extraction buffer per g tissue).
2. Centrifuge the extract at 15000 x g for 15 minutes.
3. Use the supernatant for GUS assay (enzyme extract).
4. Incubate the following mixture at 37°C.

Extraction buffer	960μl
Substrate	20μl
Enzyme extract	20μl

5. Withdraw 200 μl aliquotes from the reactions mixture at zero time and subsequent times (10, 20 and 60 min etc.) and terminate the reaction by adding 0.8ml of 0.2 M Na_2CO_3.
6. Fluorescence is then measured with excitation at 365 nm, emission at 455 nm.
7. Alternatively, the fluorescence can conveniently be detected by placing the tube on a long wave UV light box/torch.

Note: The fluorimeter should be calibrated with freshly prepared standards of 4-methyl umbelifersone (MU) (100nM-lmM) as given below.

3. CALIBRATION

Set 1μM MU equal to 5.0 relative fluorescence, and use the factor 10x if reading is low or factor 1×, if reading is too high. Convert the values obtained to nanomolar MU, and than to nanomoles MU, after correcting for the assayed volume. The preferred units to express GUS activity are nanomoles MU produced/minute.

REFERENCES

1. Hodal, L, Bochardt, A., Nielsen, J.E., Mattsson, O, and Okkels, F.T. 1992. Plant Sci., 87 : 115-122.
2. Hull, A.G. and Devic, M. 1995. In: Methods in Molecular Biology: Plant gene Transfer and Expression Protocols (Jones, H.ed.) Humana Press Inc, Totowa, NJ, Vol 49. : 125-141.
3. Jefferson, R.A. 1987, Plant Mol. Biol. Rep., 5 : 387-407.
4. Martin,T. Schmidt, R. Altmann, T., and Frommer, W.1992 Plant Mol. Biol. Rep, 10 : 37-46.

22

Detection of Neomycin Phosphotransferase II (NPT II) Activity in Transformed Plant Tissues

The development of plant cell transformation systems using the Ti plasmid of *Agrobacterium tumefaciens* as a vector is dependent upon the expression of chimeric genes carrying dominant selectable markers to identify transformants. One of the widely utilized selectable markers is the bacterial Tn5 neomycine phosphotransferase II (NPT II) enzyme that confers resistance to the transformed plant material were done by gel filtration accompanied by paper chromatography. Reiss *et al.* (1984) developed a sensitive, quantitative method for the detection of NPTII activity that has since been utilized extensively for the detection of transformed plants. With the need to screen great numbers of transformed plants a faster method of screening tissue for NPTII activity using less readiolabel has been developed (Mc Donnel *et.al.* 1987) dilution of γ^{32} P-ATP with "cold" nucleotide increases the amount of radioactive neomycine phosphate formed while reducing nonspecific background signals. These effects are likely achieved in several ways. Dilution of radioactive ATP with the cold nucleotide reduces the amount of isotope available for non-specific protein kinase side reactions, which lowers the background signal often seen in gel assays. Under the same conditions, NPTII appears to be limited only by amount of ATP available. Hence, adding more nucleotide leads to proportionately more product and no loss of signal. The inclusion of higher levels of cold ATP in combination with fluoride (inhibitor of phosphatases) helps to prevent hydrolysis of neomycine phosphate and ATP by phosphatases. This suppresses non-specific adsorbtion to phosphocellulose.

BUFFER SOLUTIONS

Extraction buffers

(can use either A or B, same for gel assays)

A. 1× Stock

		Final concentration
Glycerol	1ml	10%
0.5M Tris-HCl pH 6.8	1.25ml	62.25mM
β-mercaptoethanol	0.5ml	5%
SDS	10mg	0.1%
Final Volume in H_2O	10ml	

B. 2× Stock

		Final concentration
Glycerol	2ml	20%
0.5. Tris-HCl pH 6.8	2.5ml	125mM
10% Na-deoxycholate	2ml	2%
Final volume in H_2O	10ml	

Reaction buffer

		Final concentration
Tris	2.01g	335mM
$MgCl_2.6H_2O$	2.13g	210mM
NH_4Cl	5.34g	2M

Dissolve in 30ml H_2O and adjust pH to 7.1 with 1M maleic acid to final volume 50ml.

Assay mixture

Reaction buffer (above)	4.936ml
ATP (10mM stock)	5μl
γ-^{32}P-ATP (10μCi/μl stock)	1.5μl
Neomycine (22mM stock)	7μl
Total volume	5ml

PROCEDURE

1. Grind callus of leaf tissue, frozen in liquid nitrogen, with a mortar and pestle. Add 150 μl to 500μl of 1× extraction buffer (depend upon the amount of tissue available, approx. 1ml buffer/g tissue) and continue grinding.

2. Transfer the ground sample into an Eppendorf tube and vortex for 45-60 seconds. Maintain sample tubes on ice. Centrifuge for 5 minutes at 4°C in a microcentrifuge and transfer the supernatant to a new tube.
3. Mix in a Eppendorf tube 15μl of sample extract and 15μl of the assay mixture and mix well. Incubate the mixture for 30 minutes at 37°C.
4. Cut an appropriate size piece of Whatman™ P81 (cellulose phosphate) paper for the number of samples to be spotted. Soak the paper in a solution of 20mM ATP, 100mM pyrophosphate and dry. This eliminates non-specific binding to the paper and background in the radiogram.
5. After incubation is completed, centrifuge the tubes for 5 minutes at 10000 rpm. This step is especially necessary with plant extract samples that have been frozen for storage and than rethawed. These samples may be gelatinous and label can stick non-specifically to the paper causing false positives.
6. Spot 20μl of each sample supernatant onto the dried P81 paper allowing sufficient space between samples for spreading (2cm/spot). Usually non transformed plant tissue and bacterial extract controls are spotted on the paper.
7. After the spots have dried at 37°C for 5-10 minutes wash the paper for 2 minutes in 10mM sodium phosphate buffer, pH 7.5 at 80°C.
8. Wash the blot in the 10mM phosphate buffer at room temperature for 10 minutes (phosphate buffer helps to elimination background counts). Repeat this procedure 3-5 times. Dispose of all washes as radioactive waste.
9. Dry the blot and expose to X-rays films for 16 hrs. The positives will show up on the film as dark spots and be identified by placing the film over the original blot (or duplicate can be made in #6 above) and counted on a scintillation counter.

REFERENCES

1. Cabnes-Bastos, E.,Day, A.G., and Lichtenstein, L.P: 1989. Gene, 77: 169-176.
2. Curtis, I.S., Power, J,B., and Davey, M.R. 1995. In : Methods in Molecular Biology: Gene transfer and Expression Protocols (Jone.H.ed.) Humana Press Inc., Totowa NJ, Vol 49 : 149-159.
3. Reiss, B, Sprengel, R. Will, H. and Schaller. H. 1984. Gene. 30.211-218
4. Weide, R., Koornneef, M. and Zabel P. 1989. Theor, Appl. Genet. 78 : 169-172.

23

GENETIC TRANSFORMATION OF PLANTS USING *AGROBACTERIUM RHIZOGENES* VECTOR SYSTEM

In plant transformation in certain species specially monocots using *Agrobacterium rhizogenes* are advantageous compared to the more commonly used *Agrobacterium tumefaciens* techniques (1). Most plant cells transformed with *A. rhizogenes* develop into transgenic roots that are easily recognized, separated, and cultivated *in vitro* (2,3). From these root cultures a number of plants regenerate spontaneously or following hormone application (2-4), and, despite the "hairy root phenotype" (2) of regenerated plans, *A. rhizogenes* - mediated transformation is useful for recalcitrant and monocot species. To establish a useful *A. rhizogenes* transformation system, several strains should be tested for the virulence and the seyerity of the "hairy root phenotype." For *Lotus cornisculatus* (bird's-foot trefoil) the C58CI pRil5834 strain is used extensively and the *Agrobacterium* gene transfer system is exploited in three different strategies (10).

1. Binary vectors are conjugated or transformed into *A. rhizogenes* and the "binary" T-DNA co-transferred into the plant cells (4)
2. Integration vectors containing a restriction fragment from the TL-DNA are transferred together with the TL-DNA region after, homolgous recombination into the pRi plasmid (4)
3. A variant of the previous approach uses *A. rhizogenes* acceptor strains where pBR322 integrated into the TL-DNA segment is used as acceptor sequence in recombination (5-9).

For many species of the *Leguminosae* family, the combination of *in vitro* techniques for transformation and regeneration have proven difficult, and transgenic plants have only been obtained with a select group of species and varieties.

In the following sections, procedures of *A. rhizogenes*-mediated transformation and regeneration of *Lotus corniculatus* are outlined below, together with protocols for nodulating composite plants.

MATERIALS

Tissue Culture Media

1. L-S medium : Half-concentration of B5 medium without sucrose. Dissolve 1.94g of ready made B5 adjust pH to 5.5, add 12 Noble agar (Difco, Detroit, MI) and water to 1 L, then autoclave. Medium can be remelted once in a microwave oven.
2. L medium : Half-concentration B5 medium. Dissolve 1.94g of B5, 10g sucrose, adjust pH to 5.5, add 9g Noble agar, and water to 1L, then autoclave, Medium can be remelted once in a microwave oven.
3. LW medium : Half-concentration B5 with 0.01% Tween-20. Dissolve 1.94 g B5, 10g sucrose, add 100μl Tween-20, 8ml of 1M NaH_2PO_4 (pH 5.5), adjust volume to 1L, then autoclave.

Buffers

1. SP plus Tween 0.01% : For 1L dissolve 8.5g NaCl, 7.0 g K_2HPO_4, 3.0g KH_2PO_4, add 100μl Tween 20, adjust pH to 7.2, autoclave.
2. 2% hypochlorite, 0.02% Tween : For 20 ml mix 2.7 ml of 15% sodium hypochlorite solution with autoclaved water in a sterile universal container, add 4μL Tween 20, shake gently, and use on the same day.

Media for Growth of Bacteria

1. YB medium of *Rhizobium loti inoculum* : Dissolve 0.4g yeast extract, 0.5g K_2HPO_4, 0.2g $MgSO_4$. $7H_2O$, 0.1g NaCl, adjust pH to 6.8 and volume to 1L, autoclave.
2. YMA medium for *R. lotiplates* : Add 2g mannitol and 15g agar to 1L of YB medium, autoclave.
3. LB medium for *Agrobacteium* and *E. coli* : Mix 5g yeast extact, 10g bacto tryptone, 5g NaCl, adjust pH to 7.0, and volume to 1L, autoclave.
4. LA medium for *Agrobacterium* and *E. coli* plates : Add 14 g of agar to. 1L of LB before autoclaving.
5. B&D plant growth medium (9) : Five milliliters of four stock solutions, A, B, C and D are mixed with 1L of water. Stock A contains 294.1 g $CaCl_2.2H_2O$ per L. Stock B contains 136.1 g

KH_2PO_4 per L. Stock C contains 6.7g Fe-citrate per L, heat to dissove. Stock D contains; 123.3g $MgSO_4.7H_2O$, 87.0g K_2SO_4, 0.338g $MnSO_4.H_2O$, 0.247g H_3BO_3, 0.288g. $ZnSO_4.7H_2O$, 0.1g $CuSO_4.5H_2O$, 0.056g $CoSO_4.7H_2O$, 0.048g $Na_2MoO_4.2H_2O$.per L.Stocks are stored in brown bottles at 4°C. To avoid precipitation of B&D medium autoclave B separately and add to medium afterward.

Antibiotics and Fungicides

1. Cefotaxime at 50mg/mL : For 1L medium, dissolve 500mg in 10mL H_2O filter sterilize, and use on the same day.
2. Kanamycin at 100mg/mL : Dissolve 3.2g in 25mL H_2O, filter sterilizc. Store in aliquotes at 20°C.
3. Rifampicin at 50mg/ml; Dissolve lg of rimactan in 20 mL of H_2O; filter sterilize, store in aliquots at -20°C.
4. Spectinomycin at 100 mg/ml. Dissolve 2g of spectinomycin in 20 mL of H_2O, filter sterilize, store in aliquots at -20°C.
5. Carbendazime at 30 mg/ml: Dissolve 300 mg carbendazime in 10 mL of DMSO, store in aliquots at -20°C.

Miscellaneous

Seeds of *L. conrniculantus*,
Pouches,
Nitrocellulose filters
Universal containers
Hypodermic needles,
Carbendazime.

Methods

TRIPARENTAL MATING OF BINARY VECTORS INTO AGROBACTERIUM

1. Streak the *E. coli* strain containing the binary vector and the *E. coli* strain carrying the helper plasmid (s) onto fresh selective LA plates Incubate overnight at 37°C.
2. Inoculate 5 mL of liquid medium (LB) with *Agrobacterium* at high denstity, grow overnight at 28°C without selection.
3. Start liquid culture (LB) of each of the E. coli strains containing binary vector and the helper plasmid(s). Scrape from selective plates using a toothpick, inoculate at high density, and grow at 37°C without selection for 1 h.

4. Arrange nitrocellulose filters on well dried solid LA plates and spot 50μl of the *Agrobacterium* overnight culture onto the mating filters. Spot 200μl of an *E. coli* culture containing a binary vector and 200μl of the *E. coli* carrying helper plasmid(s) on top of the *Agrobacterium.* As controls, spot *Agrobacterium* and the *E. coli* strains on separate filters.
5. Leave the plates in a vertical flow bench, without lids, until the liquid has dried from the filters, incubate at 28°C overnight.
6. Use forceps to move the filters into universal containers with 4ml of SP Buffer. Tighten the lids and vortex until the bacteria are washed off the filter and form a suspension in the buffer.
7. Plate 10-20μl of the bacterial suspension onto LA plates selecting for the *A. rhizongenes* C58C1 chromosomal rifampicin resistance marker and the marker on the binary vector. Incubate at 28°C for 2 days.
8. Restreak transconjugants on new selective LA plates before use rifampicin and kanamycin.

CONJUGATIONAL TRANSFER OF INTEGRATION VECTORS INTO *AGROBACTERIUM*

1. In this procedure the *E. coli* containing the integration vector and the *E. coli* carrying the helper plasmid(s) are crossed before mating to the *Agrobacterium* strain. Scrape the strains from selective plates and start 3ml liquid cultures with high inoculum, incubate at 37°C for 3h. Start the cross by transferring 1.5 ml of a *rec*A *E. coli* containing the integration vector and 0.3 ml of *E. coli* carrying helper plasmid (s) to the same tube. Incubate for 1h at 37°C without shaking. Plate 100μl from the cross and each of the controls on LA plates selecting for integration vector and helper plasmid(s). Incubate overnight at 37°C.
2. The colonies appearing after selection for both integration vector and helper plasmid(s) are used to start the *E. coli* donor culture for the *Agrobacterium* mating. The procedure is the filter procedure used for mating of binary vectors except that 200μl of *E. coli* culture and 50μl *Agrobacterium* is used for the mating, and that 50 to 100 μl are plated onto selective LA plates from the SP buffer suspensions.

Transformation Protocol

1. Take lids off individual 9 cm Petri dishes and arrange the dishes at an angle using the lids as a support. Cast the agar slopes

pouring approx 20 ml of L-S medium into the lowest part of the plates.

2. Surface sterilize seeds in 20ml of 2% hypochlorite, 0.02% Tween. Shake for 20 min and rinse the seeds in 6 × 20ml of autoclaved water. Transfer the seeds onto a sterile filter paper in a Petri dish, add 3-5 mL of sterile water and incubate at 26°C overnight.
3. Arrange 10-12 seeds at the top of the agar slope in the LS plates. Pack 10 plates together using transparent tape. Incubate plates at 26°C continuous light for 5 days.
4. Streak the *Agrobacterium rhizogenes* strains for the transformation onto fresh selective medium on day 3 after the start of seed germination. Plates for histidine auxotrophic strains are supplemented with 100% μg/ml of histidine. Incubate at 28°C.
5. Seedlings are "wound-site infected on day 5 after germination using a hypodermic needle. A needle is fitted onto a 2 ml syringe and dipped into a freshly grown *Agrobacterium* colony, on a selective plate. The hypocotyls are then wounded in two positions by gently stabbing the needle into the plant tissue. Control plants are wounded without bacteria.
6. Pack the plates together again and keep wounded plants overnight in the dark at 21°C continuous light for 2-3 weeks to allow hairy roots to develop from the wound sites.
7. Stem fragments with emerging hairy roots are transferred to L medium containing cefotaxime at 300-500 μg/mL. To establish the root cultures effectively, transfer 1-2 cm of stem on either side of the wound site. When binary vectors T-DNA, for example, Kanamycin at 50μg/ml for the pGV941 T-DNA.
8. Incubate in the dark at 26°C for 2 weeks.
9. Transfer individual roots longer than 3 cm, separately, to L medium containing cefotaxime at 300-500 μg/ml. Incubate in the dark for 2 or 3 weeks. If binary vectors were used, continue selecting for the marker on the binary T-DNA.
10. Remove agar from root system and transfer as much as possible of the roots onto L medium containing cefotaxime at 300-500 μg/mL. Incubate in the dark for 1 wk or longer Continue selecting for the binary T-DNA.
11. Transfer established root cultures from step 4 or step 3 above into continuous light [philips TLD (Einhoven Holland) 30 W/33tube]

at 26°C. A small part of the root culture is subcultured on L medium with selection and kept in the dark as root culture.

12. Leave the root cultures in the light without changing the medium. Spontaneous shoot formation is first visible as green spots on the roots. Shoots will then develop from the spots and continue to grow into proper elongated shoots.
13. Shoots of 3-4 cm in lengths are cut off, if possible together with a bit of root, and rooted for 2 weeks on100mL of L medium in plant containers. Kanamycin at 50μg/mL can be used as extra selection for binary T-DNA.
14. Transfer plants to pots containing a nitrogen free support, water with B&D medium supplemented with 5mM KNO_3 and inoculate with 2 mL of *Rhizobium loti* culture grown for 3-4 days in YB medium. Cover with a plastic bag.
15. Remove plastic bags to harden the plants continue to water with B&D.

Note: For selection of *Agrobacterium* C58C1 pRi 15834 and for derived strain, use the following antibiotics with concentrations as : Rifampicin (100 μg/ml), Kanamycin (300 μg/ml), Spectinomycin (100 μg/ml), Streptomycin (100 μg/ml), Gentamycin (100 μg/ml) and Tetracycline (l-2 μg/ml).

REFERENCES

1. Broughton, W.J . and Dilworth, M. (1971). Biochem. J. 125, 1075-1080.
2. Handberg, K. and Stougaard, J. (1992) Plant J. 2, 487-496.
3. Hansen, J. Jorgnsen, J.E. Stougaard, J., and Marcker, K.A. (1989). Plant Cell Rep. 8, 12-15.
4. Mugnier, J. (1988). Plant Cell Rep. 7, 9-12.
5. Stougaard, J. Abidstem. D. and Marcker , K.A. (1987). Mol. Gen. Genet 207, 251-255.
6. Schmulling, T, Fladung, M, Grossmann, K, and schell, J. (1993). Plant, J. 3, 371-3421.
7. Shen, W.H. Petit, A Guren , J., and Tempe J. (1988). Proc. Natl. Acad, Sci. USA 85, 3417-3421.
8. Stougaard, J. 1995. *In:* Methods in Molecular Biology-Plant Gene Transformed Expression Protocols (Jones, H. ed.) Humana Press Inc. Totowa, NJ. Vol. 49:49-61.
9. Tepfer, D. (1984) Cell 37, 959 - 967.
10. Vilaine, F. and Casse-Delbart, F. (1987). Gene, 55 : 105 - 114.

24

TOBACCO CHLOROPLAST TRANSFORMATION USING MICROPROJECTILE BOMBARDMENTS (BIOLISTIC GUN)

The ultimate goal of plant genetic engineering is to produce novel transgenic plants with improved agronomic traits. The production of transgenic plants depends among other factors on the stable introduction of foreign gene into the plant genome. A large number of gene transfer methods are currently available . However, the method of gene transfer using the Microprojectile/Particle Delivery System (PDS-1000) is most versatile and has proved successful in introducing foreign genes not only to the nucleus but also to the chloroplasts in plants (Fig. 24.1). The protocol described below has been standardised for tobacco chloroplast transformation using tungsten particles.

I. Preparation of Microprojectiles

1. Use 50 mg of tungsten powder (BioRad Tungsten M-17) for 100 shots.
2. Add 1ml of freshly-opened absolute ethanol.
3. Centrifuge for 10 secs in microcentrifuge and keep in water bath 95°C for 1 hr.
4. Replace with 1ml absolute ethanol and centrifuge 10 secs. Can be frozen O/N at -20°C.
5. Remove ethanol. Add 1 ml. water, resuspend tungsten and recentrifuge. Repeate 3 times.
6. Finally resuspend tungsten in 1 ml water.
7. Sample preparation for 6 shots : (important - add in order)
   - 50 μl tungsten solution
   - 10 μl DNA in TE buffer (atleast 1 μg/μl).
   - 50μl 2.5M $CaCl_2$

- 20 µl 0.1M spermidine (free base.)

8. Add 200µl ethanol
9. Centrifuge for 10 secs in a micorofuge.
10. Remove supernatant Add 200 µl ethanol and resuspend extremely well, making sure all clumps are dispersed. Repeat 3 times.
11. Resuspend final pellet in 30 µl ethanol.
12. Put 5µl onto each disk.

II. Preparation of leaf samples for shooting

Each shot requires a target leaf, and for one construct, you need to cut 20-30 leaves This number of leaves can easily be obtained from 8 magenta boxes of cuttings. For cutting, use sterile forceps and scalpels. Each cut leaf is placed abaxial side up on top of a sterile filter disc (Whatman #4 circles) on MS plate. Try to centre the leaf into the plate, to maximize the surface receiving the shot. The leaf surface should touch the MS medium as much as possible. Therefore, if the leaf is slightly curled up, nich it to allow it to lie flat. Leaves should be cut the day before you plant to shoot. A stack of thin plates containing cut leaves is bagged in plastic and left in the culture room overnight.

III. Shooting leaf samples

For each shot you will need one 1100 p.s.i. rupture disc (the small dark brown discs), one flying disc (the large orange discs), and one stopping screen (the wire mesh screens). All these are supplied in the Bioard shooting kit (500' Standard pressure kit Before shooting sterilize the number of of discs and stopping screens that you will need by placing them for 5-10 minutes in a petri dish containing absolute ethanol. The discs and screens are then dried by standing them up along the side of a sterile petri dish and allowing them to air-dry under sterile conditions in the tissue culture hood.

While sterilizing the disposable shooting components, the gun parts should also be surface sterilized by wiping with absolute ethanol. Three components can be removed from the gun vacuum chamber for surface sterilization. These are the holder for the rupture discs, the stage and holder assembly for the flying discs, and the platform for the leaf samples. After removing these and wiping with ethanol, the interior of the vacuum chamber and door should be wiped with absolute ethanol. Finally the five metal ring holders for the flying discs should be surface sterilized.

Now you are ready to shoot.

1. Turn the gun on (first red button on the left)

   Turn the pump on.

   Turn the gas cylinder on. First, turn the screw bar counter clockwise until it feels loose. Then turn the knob on top of the cylinder counter counterlock wise (the pressure valve on the right should read at least 1300 p.s.i.). Then turn the screw bar slowly clockwise until the pressure valve on the right should read at least 1300 p.s.i.) Then turn the screw bar slowly clockwise until the pressure valve on the left reads 1100 p.s.i.
2. Place the five flying disc holders in a petri plate and into each one place a flying disc (the large orange ones.) Press down the discs with the red plastic "capplug", so that they fit snugly into the holders.
3. Onto the surface of each disc, pipette 5μl of DNA-coated tungsten.
4. You are now ready to set up a shot. Do all the following steps in the same order for each shot:

   (i) Place a rupture disc into the rupture disc holder, using forceps. Make sure the disc is properly seated in the holder by tapping the holder a few times gently. Screw the holder into the vacuum chamber. It must be screwed in tightly.

   (ii) Place a stopping screen onto the diaphragm in the flying disc holder assembly. On top of this, place UPSIDE DOWN a metal ring containing a flying disc with DNA on it. Secure all the components by screwing on the assembly ring. Place the stage in the second slot from the top in the vacuum chamber.

   (iii) Take a petri plate with a leaf sample and place it on the sample platform (lid off.). The sample platform should be positioned in the chamber in the fifth slot from the top (second from the bottom).

   (iv) Press the vacuum button (middle red button, up position to begin pumping air out of the shooting chamber. Allow the vacuum pressure to reach at least 25 inches of Hg. Then press the shooting button CONTINUOUSLY (red button on the right, up position). When the pressure reaches 1100 p.s.i. the rupature disc inside the gun should break (you will hear a pop, and see the petri plate jump.) Release the shooting button, and remove the vacuum by setting the vacuum button to vent (middle position).

(iv) Take the petri plate out of the chamber. Repeat until all samples have been processed.

IV GUS Histochemistry to test the above method

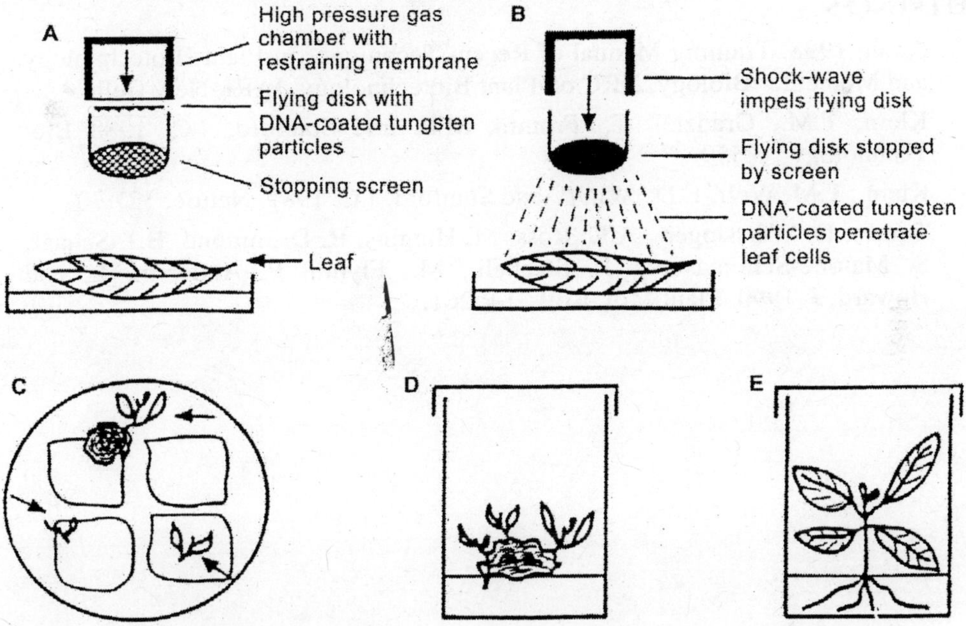

Fig. 24.1 Plastid transformation in tobacco. Different steps (A to E) are shown. **A.** Preparation of leaves for shoting; **B.** Shooting plated leaves with DNA coated tungsten particles using Biolistic Gun; **C.** Cutting leaves into segments and plating on selection medium containing spectinomycin; **D.** Regeneration of shoots on selection medium, and E\ Rooting of the regenerated shoots.

Leaves can be shot with a *uid* A gene engineered for nuclear expression from the 35S promoter. Transient expression of this construct can be monitored in shot leaves, for GUS activity (see below). If the tungsten coating and shooting conditions were optimum, you should see ~ 100 blue spots on the shot leaf surface.

X-GLUC solution (20ml):

10mg X-gluc in 100μl dimethylformamide.

1ml 1 M sodium phosphate pH 7.0

1ml Triton (from a 20% stock solution)

18 ml H_2O.

Incubate leaf in the above solution for 24-48 h at 37°C in dark.

Observe GUS positive blue spots on the leaf, and section the leaf if required to determine the histochemical localization of GUS activity.

REFERENCES

1. Anon, 1998. Training Manual of Recent Techniques in Plant Biotechnology and Molecular Biology. NRC on Plant Biotechnology, IARI, New Delhi.
2. Klern, T.M., Gradziel, T., Fromm, M.E. and Stanford, J.C. 1988 Bio/ Technology , 6:559
3. Klein , T.M. Wolf, E.D., Wu, R. and Stanford, J.C. 1987. Nature, 327:70
4. Tomes, D. Weissinger , A.K. Ross, M. Higgins, R. Drimmond, B.J, Schaaf, S. Malone-Schomasberg J., Staebell,, M., Flynm, P. Andersm, J., and Howard, J. 1990. Plant Mol. Brol., 14: 261.

25

MICROPROJECTILE MEDIATED TRANSFORMATION OF RICE USING BIOLISTIC GUN

The following procedure has been standardized for the basmati rice (Basmati 370 and Karnal Local and also found to work fairly well with the elite indica rice cultivars. The procedure involves use of mature embryos as explants.

I. Preparation of the target material

1. Use seeds obtained from a pure line plot [stored at room temperature (20°-30°C) for not more than a year]. Select out uniform looking and apparently healthy looking for further processing.
2. Manually dehull the grains taking care that the embryo region remains undamaged.
3. Screen dehulled grains under a dissecting microscope; remove all those harbouring any fungal colonies, as also shrivelled and grains of smaller size.
4. All further operations should be carried out under a laminar flow hood. Surface sterilize seeds by treating with absolute alcohol for a minute accompanied with some vigorous shaking. Remove ethanol and wash 3-4 times with sterilized distilled water. Treat the seeds now with 0.1% $HgCl_2$ (With a drop of Tween-20 to act as a surfactant) for about 8-10 minutes. Ensure that the treatment is carried out with vigorous shaking manually or on a shaker.) After this treated seeds are washed thoroughly, atleast 5-6 times with sterile distilled water.
5. Drain put all water except a little bit to keep to seeds wet. Lift single seeds with a forceps and plate them on the callusing medium (Murashige and Skoog's basal medium supplemented with 2,4-D at 2mg/L, 3% sucrose and gelled with 3% Gelrite). As many as 10-15 seeds may be plated in a 60mm Petri-dish. Plated seeds are incubated in dark at 25°C for 4-5 days.

6. Isolate scutellar callus, taking care that the process involves minimal damage to the callus. Transfer such calli to a fresh medium in a 60 mm dish. Arrange 20-25 calli within about 1.5 cm diameter area in the centre of the dish. The embryo calli are ready for bombardment.

II. Preparation of particles for bombardment

1. Materials and solutions
   * Autocalved 0.5 ml and 1.5 ml Eppendorf tubes.
   * Filter sterilised 70% EtOH and EtOH absolute (abs).
   * Sterilised DNA solution : Precipitate DNA overnight in a 1.5 ml Eppendorf tube by adding 1/10 vol. of 4M LiCl and 3 vol. of EtOH abs. Centrifuge for 10 min. at room temperature at a full speed (14,0000 rpm), wash the DNA pellet with 1ml filter sterilised 70% EtOH and repeat centrifugation. Let the pellet dry in a clean laminar flow bench and dissolve it in sterile ddH_2O at a final concentration of 1 µg/µl.
   * Gold particles : Weigh 50 mg gold particles and wash once with sterilised distilled water. Add 1 ml of 50% glycerol. Autoclave in punctured 1.5 ml Eppendorf tubes. After autoclaving, the mixture is transferred to new Eppendorf tubes (unpunctured) and stored at RT.
   * 2.5M filter sterilised $CaCl_2$ solution
   * 100 mM filter sterilised Spermidine solution

III. Procedure for particle preparation

1. Vortex the gold-glycerol suspension for about a min.
2. Pipette out an aliquot of 50 µl in a 0.5 ml autoclaved Eppendorf tube. Add DNA (5 µg) and vortex for about 30 secs.
3. Add 50 µl of 2.5M $CaCl_2$ solution.
4. Add 20 µl of 100 mM Supermidine solution.
5. Vortex for about 3 mins. at max. speed.
6. Keep the tube on ice for about 5 mins. and then spin down by short-pulse centrifugation.
7. In a laminar flow hood, remove supernatant and wash 2x with filter sterilised EtOH abs.
8. add 50µl of sterile EtOH abs.

Note : * This amount of particles is sufficient for 5 bombardments.

* Prepare the particles on the day of shooting.
* Take care not to prolong the ice treatment at step 6 to avoid the DNA coated particles to get clumpy.

REFERENCES

1. Anon, 1998. Training Manual on Recent Techniques in Plant Biotechnology and Molecular Biology. NRC on Plant Biotechnology, IARI, New Delhi.
2. Cao, J, Duan, X., McElroy, D. and Wu, R.1992. Plant Cell Reports, 11 : 586-591.
3. Klein, T.M. Roth, B.A. and Fromm. E.M. 1989. *In:* Genetic Engineering, Principles and Methods II (L. Setlov, J.K. ed) Plenum, New York. pp. 13.

26

INOCULATION AND STORAGE OF *ESCHERICHIA COLI* FROM OVERNIGHT CULTURE

Engineered *E. coli* strain (DH5α), derived originally from MM 291 crossed with KL 16-99, are used for cloning and for making genomic/c DNA libraries using plasmid/cosmids as vectors. The derivative of that cross is DH1 which was modified to DH5 and then to DH5α. Many prokaryotic and eukaryotic genes could be expressed. The engineered *E. coli* strains contain defective galactosidase gene coding for a factor which is complemented by the aminoterminal portion of the defective galactosidase (a peptide), coded by a DNA fragment present in the plasmid. This complementation produces functional tetrameric enzyme galactosidase which can detach galactose from the coloured dye X-gal to produce blue colour colonies by bacteria containing plasmids without an insert (X-gal is rendered colourless due to association of galactose).

BUFFERS AND SOLUTIONS

Sterile LB media; Ampicillin (100 mg/ml) solution; Overnight grown culture; LB agar plates; Sterile loop; Sterile glycerol.

Protocol

A. Inoculation from liquid culture

1. Add 50 ml LB medium into each of two sterile 250 ml conical flasks.
2. Add 30 μl of 100 mg/ml ampicillin solution in one of the two flasks.
3. Add 500 μl of previous overnight grown culture in each of the two flasks.
4. Incubate them in shaker at 37°C overnight.

B. Inoculation of single colony from an agar plate

1. Add 2 ml LB medium into each of two 10 ml sterile tubes.
2. Add 1.2 µl of 100 mg/ml ampicillin solution in one tube.
3. With a sterile loop lift a single colony.
4. Immerse in LB with ampicillin, shake cells off; repeat the step for tube without ampicillin.
5. Incubate in shaker bath at 37°C overnight.

C. Storage of cultured bacteria

1. Add 150 µl sterile glycerol to a sterile storage tube.
2. Add to it 850 µl of an overnight grown culture of the bacteria.
3. Vortex for 1 min.
4. Store at -20°C for no more than 4 months; storage at -70°C is good for long time.

Determination of Growth curve

BUFFERS AND REAGENTS

Sterile LB medium; Overnight grown culture of DH5α; Sterile loop.

Protocol

1. Inoculate twenty vials (numbering 1 to 20) containing fresh LB culture medium (10 ml each) with 100 µl of overnight grown culture of DH5α.
2. Incubate them at 37°C with constant shaking in the laboratory shaker (200 rpm).
3. Take out the culture vials number 1 & 2 after one hour and 30 min respectively and keep on ice.
4. Take out the remaining culture vials (no. 3 to 20) sequentially after every 15 min and keep on ice.
5. Finally measure the OD_{550} of each culture and note.
6. Plot a graph using the data obtained and study Lag phase, Log phase and Stationary phase.

SINGLE CELL CULTURE (STREAK-PLATE METHOD)

BUFFERS AND REAGENTS

Sterile LB medium; Overnight grown culture of DH5α; Sterile loop; LB agar plates.

Protocol

1. Flame a wire loop which is about 3 mm across and has a stem of about 6 cm, allow the loop to cool, or cool by immersing in a sterile area of agar.
2. Flame the mouth of the tube with overnight broth culture of DH5α cells and remove a loopful of cells.
3. Streak the cells at one side of a well-dried LB-agar plate at position 1 (see Fig.26.1); streak several times close together.
4. Flame the loop and cool carefully at one side of the plate (position L).
5. Streak again at position 2 shown in the figure.
6. Flame the loop and cool as before. Repeat steps 4 and 5 as indicated at positions 3, 4, 5, 6 in the figure.
7. Seal the Petri plate with parafilm.
8. Incubate the plate at 37°C with the agar facing downwards to minimize contamination and to reduce the chance of droplets of condensation falling on the agar surface.

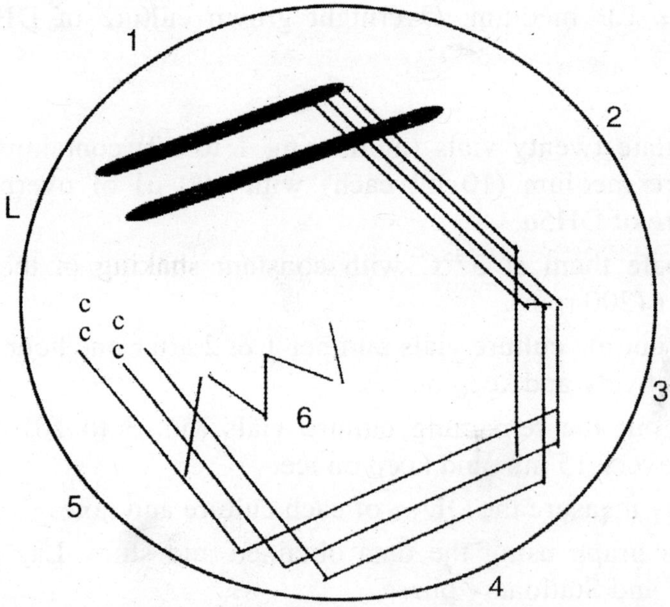

Fig. 26.1

REFERENCES

1. Anon. 1996. Training Manual on Molecular Biology and Crop Biotechnology. CCS University, Meerut (India)

Construction of Genomic DNA Library

Genomic libraries or other DNA libraries are constructed with the objective of identifying genes and for subsequent detailed analysis of gene organization, function and regulation. The genomic library normally consists of a vector with fragments of DNA cloned into one of the restriction sites of the vector. The fragments of the genomic DNA are generated by partial digestion by restriction enzyme. This is done to ensure fragment carrying complete gene are generated. The size of the fragments cloned is dependent on the nature of the vector used. Genomic libraries can be variously screened and clone(s) correspon-ding to various genes identified. One of the easiest approaches is to identify other clone(s) among the library capable of complementing functionally the mutant gene in the recipient through triparental bacterial matings. Genomic libraries are thus of immense value in identifying and analyzing genes in an organism.

A. Reagents, Solutions & Buffers

1. **STE (10×)**

   10 mM Tris HCl (pH 8.0)

   100 mM NaCl

   1 mM EDTA (pH 8.0)

2. **SM Buffer (per liter)**

   5.8 g of NaCl

   2.0 g of $MgSO_4.7H_2O$

   50 ml of 1M Tris HCl (pH 7.5)

   5 ml of 2% (w/v) gelatin

   Adjust final volume to 1 litre and autoclave.

3. **P1 Buffer**

   Tris HCl 50mM

EDTA	10mM
RNase	100 mg/ml
pH	8.0

P2 Buffer

NaOH	200 mM
SDS	1%

P3 Buffer

Potassium acetate	3 mM
pH	5.5

4. **QBT Buffer**

NaCl	750 mM.
MOPS	50 mM
Ethanol	15%
Triton × 100	0.15%
pH	7.0

QC Buffer

NaCl	1M
MOPS	50 mM
Ethanol	15%
pH	7.0

QF Buffer

NaCl	1.25 M
Tris-HCl	50 mM
Ethanol	0.15%
pH	8.5

5. **Solutions for Plasmid DNA Isolation (mini-prep)**

Solution A

Tris HCl	25 mM
pH	8.0

Solution B

NaOH	0.2 N
SDS	1%

Solution C

Ammonium acetate	7.5 M

6. **Gel Electrophoresis buffer**

TAE	50X
Tris base	242 g
Glacial acetic acid	57.1 ml
0.5 M EDTA (pH 8.0)	100 ml

TE Buffer

Tris	10 mM
EDTA	1 mM

B. 1. **LB (Luria Broth) Media**

NaCl	10 g
Tryptone	10g
Yeast extract	5g

Distilled H_2O to make 1 litre

16 g agar (to make solid medium)

2. **TYM**

Bacto tryptone	5g
Yeast extract	0.5 g
Mannitol	10g
$CaCl_2$ (lM)	1ml
Agar	16 g

Distilled water to make 1 litre.

C. **Antibiotics**

Tetracycline (10 mg/ml) in ethanol. Filter sterilized, store in dark tube at -20°C.

PREPARATION OF AGAROSE GEL AND DNA FRACTIONATION BY ELECTROPHORESIS

Buffers and Reagents

Agarose, 10× TBE buffer (0.9 M Tris-Base, 0.9 M boric acid and 0.02 M EDTA); 10 mg/ml ethidium bromide; Gel casting assembly; Gel loading buffer.

Protocol

1. Prepare a 0.7% (w/v) agarose gel suspension by heating in 1× TBE buffer. The solution shall be kept as liquid by keeping it at 45°C.
2. Set the electrophoresis apparatus by joining + & – electrodes with electric current wire.

3. Cast the gel in a tray, kept on horizontal surface.
4. Place a comb in such a way that 2mm gap is maintained between the bottom of the gel and the comb tip.
5. Allow the gel to solidify.
6. Place the gel in the chamber and add 1× TBE buffer to fill the chamber and flood the surface of the gel.
7. Carefully remove the comb (just prior to use of the gel).
8. Connect the apparatus with the power pack.
9. Load the DNA (with the gel loading buffer) within the slots.
10. Allow the DNA to run through the gel at 2-5V/cm, for 2-8 hours.
11. After electrophoresis, stain the gel (if the ethidium bromide was not added to the gel and/or in the buffer) in 0.5 μg/ml ethidium bromide for 20-30 min at room temperature.
12. Observe and photograph the DNA under UV illumination using the GDS.

1. PREPARATION OF (pUC 18 PLASMID) VECTOR DNA

1. Centrifuge overnight grown 100ml culture of DH 5α (pLAFR1) at 15000g for 10 minutes. Discard the supernatant.
2. Respond the bacterial pellet in 4ml of buffer P_1.
3. Add 4 ml of buffer P_2, mix by inversion and incubate at room temperature for 5 minutes.
4. Add 4 ml of chilled buffer P_3, mix gently and incubate on ice for 15 minutes.
5. Centrifuge at 15000g for 30 minutes at 4°C. Remove supernatant promptly.
6. Equilibrate a Qiagen Tip 100 column with 4 ml of buffer QBT.
7. Apply supernatant of Step 5 to the tip and allow it to empty by gravity flow.
8. Wash the tip 2 × with 10 ml buffer QC.
9. Elute DNA with 5 ml buffer QF.
10. Precipitate DNA with 0.7 volume of isopropanol. Centrifuge at 15000 × g for 30 minutes at 4°C and remove the supernatant.
11. Wash DNA with 2 ml of 70% ethanol, air dry for 5 minutes and redissolve in 100 μl of TE buffer.

12. For linearizing, digest the DNA with Eco R1 (@ 10 U/μg of DNA) at 37°C for 3 hours.
13. Check an aliquot on 0.7% agaorse gel for complete linearization using DH5α as control in a parallel slot.
14. For dephosphorylation, treat the restricted DH5α DNA with alkaline phosphatase (CIAP) @ 0.05 U/μg for 1 hours.
15. Inactivate remaining CLAP enzyme using proteinase K (50 μg/ml) in STE buffer having 0.5% SDS. Incubate at 65°C for 1 hour.
16. Clean the digest mix with Gene clean Kit.

2. PREPARATION OF GENOMIC DNA (*Cicer-Rhizobium* COBE 13)

1. Take 1.5 ml culture *Rhizobium meliloti* strain, Rmd 201 and centrifuge at 13000 rpm for 1 minute. Remove supernatant.
2. Add 600μl nuclei lysis solution, resuspend and incubate at 80°C for 5 minutes.
3. Add 3 μl of RNase solution and mix by gentle inversion.
4. Incubate at 37°C for 60 minutes.
5. Cool and add 200 μl protein precipitation solution. Vortex at high speed.
6. Keep on ice for 5 minutes. Centrifuge at 13000 rpm for 3 minutes.
7. Transfer supernatant to a microcentrifuge tube containing 600 μl isopropanol, mix by inversion (thread like strands of DNA form visible mass).
8. Centrifuge at 13000 rpm for 2 minutes and pour out the supernatant.
9. Add 600 μl 70% ethanol, invert several times (to obtain cleaner DNA pellet). Centrifuge at 13000 rpm for 2 minutes. Pour off ethanol.
10. Air dry pellet for 20 minutes. Add 100 μl of DNA rehydration solution, incubate at 65°C for 1 hour with periodical mixing.
11. Store at 4°C.

3. RESTRICTION DIGESTION OF DNA

Protocol

1. Standardize the reaction needed for partial digestion with Eco R1 to obtain large amount of restricted fragment in the range of 17-23 kb as follows:

2. Set up restriction digestion as indicated below:

Total DNA	:	2μg	2μg	2μg	2μg
Reaction buffer (10X)	:	2μl	2μl	2μl	2μl
Eco R1	:	1U/μl	0.5U/μl	0.25U/μl	0.125 Uμl

Add sterile double distilled water to make final volume 20μl

3. Incubate the mixture for ½ hour at 37°C. Thereafter stop the reaction by maintaining the reaction mixture at 65°C for 10 minutes. Then allow it to slowly cool at room temperature for 30 minutes.
4. Load an aliquot on 0.4% agarose gel with MW marker and run at 20V till the dye reaches bottom of the gel.
5. Check the concentration of Eco RI that gives maximum amount of fragments at 17-23 kb size of DNA.
6. Carry out the same procedure with 250 μg of DNA (scaled up amount).
7. Check the large scale restricted DNA for proper digestion and isolate the required fragment size DNA (17-23 kb) and purify using gel elution kit.

4. SUCROSE GRADIENT CENTRIFUGATION

Protocol

1. Prepare 10% and 40% sucrose solution (w/v) in buffer having 1 M NaCl, 20 mM Tris Cl (pH 8.0) 5 mM EDTA.
2. Make a gradient of 10% to 40% sucrose in polyallomer tubes.
3. Load about 250 μg DNA sample by overlaying slowly along the wall.
4. Ultracentrifuge the polyallomer tubes at 24000 rpm for 26 hours at 20°C.
5. After centrifugation, place a cellophane tape at the bottom of the polyallomer tube and pierce slowly using 18 gauge needle and collect fractions of 0.5 ml volume in microfuge tubes.
6. Screen the alternate fractions on 0.4% agarose gel for locating 17-23 kb fragments using lambda Hind III digest as marker.
7. Pool fraction that contain the required size DNA fragments, purify the fractionated DNA fragments.

8. Dialyze the pooled sample for 24 hours against TE buffer at room temperature with 3 to 4 changes of the buffer.
9. Extract with 2 volumes of isobutanol 3 times and discard the top layer each time with Pasteur pipette.
10. Precipitate the DNA with double the volume of cold ethanol by keeping in ice for 2 hours.
11. Precipitate the DNA and dissolve in 100 µl TE.
12. Load 2µl of the sample on gel and check the DNA fragment size distribution.

5. LIGATION OF GENOMIC DNA FRAGMENTS ON pUC 18 (DH5α) VECTOR

Protocol

1. Mix on ice, 1 µg of dephosphorylated pUC 18 (DH5a) and lµg DNA fragments (17-23kb) in a total reaction volume of 20 µl containing 4 units of T4 DNA ligase in ligation buffer containing 1 mM ATP and adding double distilled water or Milli-Q water.
2. Incubate the reaction mixture at 15°C overnight and store at -20°C till use for packaging.
3. Run on 0.8% agarose gel, the dephosphorylated DNA inserts and ligated sample in parallel slots for test of ligation.

6. PACKAGING OF LIGATED MOLECULES (USING PACKAGING KIT)

Protocol

1. Shift from-80°C the sonic-extract (yellow tube) to ice and thaw on ice (at least 1 hour before use).
2. Thaw the frozen extract (red tube) between your fingers until the extract just begins to thaw.
3. Add the ligated DNA (1-4 µl containing 0.1-0.5 µg) to the red tube. Mix and place on ice.
4. Immediately add 15 µl of sonic extract from yellow tube to the red tube containing the DNA.
5. Stir the tube with pipette tip well without introducing any bubbles.
6. Incubate the reaction at room temperature (22°C) for 2 hours.
7. Add 500 µl SM buffer to the tubes.
8. Add 20 µl chloroform and mix the contents of tube gently.
9. Spin the tubes and transfer the supernatant into a fresh Eppendorf.

7. TRANSDUCTION OF STRAIN HB101 WITH THE PACKAGED DNA

Protocol

1. Spin overnight culture of HB 101 at 4000 rpm for 10 minutes at 4°C and gently resuspend the pellet in 12.5 ml of 10mM $MgSO_4$.
2. Adjust the OD_{600} of the suspension to 0.5 in 10mM $MgSO_4$.
3. Prepare 1:10 dilution of the packaged cosmid library in SM buffer.
4. Mix 25µl of the dilution to 25µl of HB 101 cells and incubate the tube at room temperature for 30 minutes.
5. Add 200µl LB to the above mix and incubate for 1 hour at 37°C shaking the tube gently once every 15 minutes.
6. Spin in the microfuge for 1 minute and resuspend the pellet in 50µl of fresh LB and spread on LA plate supplemented with tetracycline (@ 10 µg./ml).
7. Incubate the plates for 12 hours at 37°C.
8. Pool the colonies appearing in LA+tet plates. This pool constitutes the library and is stored at -70°C with glycerol (15-20%) for subsequent use.

8. ISOLATION OF RECOMBINANT DNA

Protocol

1. From LA+tet plates pick up twenty random colonies. Grow them in LB+tet (10 ml) as independent cultures.
2. To make DNA mini-preps, log phase *E. coli* (HB 101) cells after washing are suspended in 100 µl of Solution A and gently vortexed.
3. 100 µl of the above mix is taken in fresh eppendorf tube, 20 µl of freshly made solution B is added and mixed vigorously and incubated on ice for 20 minutes and then centrifuged at 12000 rpm for 3 minutes in microfuge.
4. Supernatant is then transferred into fresh eppendorf tubes having 1 ml of ice cold ethanol maintained at -20°C. It is subsequently mixed by inverting the tube and incubating on ice for 20 minutes.
5. It is centrifuged at 12000 rpm for 3 minutes followed by aspirating of the supernatant.
6. The tube is allowed to dry at room temperature for 10 minutes and then the DNA pellet is dissolved in 100 µl of sterile water and incubated at room temperature for 10 minutes.

7. Then 50 μl of Solution C and 300 ml of cold ethanol are added to solution of DNA, mixed well followed by incubation on ice for 15 minutes.
8. It is centrifuged for 3 minutes at 12000 rpm and supernatant is aspirated using vacuum.
9. DNA pellet is dried at room temperature for 10 minutes and dissolved in 25μl of sterile distilled water.
10. Digest plasmid DNA (1μg) with 1μl of restriction enzyme (in this case Eco R1, 12 units/μl) and 1μl of assay buffer is added.
11. The restriction reaction is carried out at 37°C for 3 hours for complete digestion.
12. Reaction is terminated by chilling on ice. The restricted DNA is loaded on 0.8% agarose gel and run at 40 volts for 15 minutes, followed by 80 volts for 2 hours. Hind III digest of lambda DNA is used as molecular weight marker.
13. Count the lanes showing the release of the insert from the cosmid vector.
14. The efficiency of the library is determined as:

$$\frac{\text{Number of clones with insert}}{\text{Total number of clones randomly picked}} \times 100$$

9. PREPARATION OF COMPETENT CELLS

The method relies on the combined effect of cold and calcium chloride to induce the cells towards competency:

Requirements

1. 2×10 ml aliquots of L-broth.
2. 1×100 ml aliquot of L-broth (#1) in a 1 litre flask, or 2×60 ml aliqotos of L-broth in two 250 ml flasks.
3. 0.1 M calcium chloride. Keep cool prior to use.
4. One 100 ml polycarbonate centrifuge tube and lid (sterile). One similar (non-sterile) tube for balancing.
5. Ice bucket.
6. Spectrophotometer.
7. Fast shaker.
8. Centrifuge.

Protocol

1. Set up 2×10 ml of the appropriate strain (JM 9) in L-broth. Commonly the cells are stored as glycerols. Use a wire loop to inoculate. Grow at 37°C on the shaking incubator.
2. Inoculate 2×60 ml aliquots of broth with 1 ml of overnight grown culture and grow at 37°C on a fast shaker until the OD_{650} reaches a value of 0.2. (this is equivalent to an O.D. of 0.1 at 600 mm for side arm flasks or an O.D. of 0.14 at 600 nm for cuvettes.
3. When the required O.D. is reached, place the cultures in ice. Transfer 100 ml of culture to a sterile 100 ml centrifuge tube. Spin down the cells at 5000 rpm using the centrifuge at 5-10°C (for 15 min).
4. Decant off the supernatant. Resuspend in 10 ml of 0.1M $CaCl_2$ solution. Keep as cool as possible. Add a further 40 ml of 0.1 M $CaCl_2$ solution.
5. Spin down the cells as for step 3. Decant off the supernatant, resuspend the pellet in 5 ml of 0.1 M $CaCl_2$ solution. Store in ice prior to use or store in 200 μl volumes at -70°C. Competent cells can be stored for six months at -70°C. The cells are used for transformation. The process described above weakens the cell to such an extent that when mixed with plasmid DNA, and given a brief incubation, a number of cells will take up the plasmid, and in doing so will be transformed into plasmid carrying hosts. These can be further screened using minipreps etc., and used to prepare large quantities of DNA for sequencing.

10. TRANSFORMATION AND PLATING

Buffers and Reagents

Ligated DNA, competent Cells, LB-medium, LB agar plates with antibiotic (ampicillin).

Protocol

1. Take 200 μl cell suspension (competent cells) in 1.5 ml EP tube placed on ice.
2. Add the ligated DNA (50-100ng) in the EP tube.
3. Leave it on ice for 30 min.
4. Transfer the tube at 42°C (water bath) and incubate for exactly 2 min without shaking.

5. Quickly transfer the tube on ice for 2 min.
6. Add 800 μl LB medium. Incubate at 37°C for 45-60 min in water bath.
7. Transfer 200 μl transformed competent cells onto LB agar plate containing the appropriate antibiotic (60 μl/ml ampicillin).
8. Leave the plate at room temperature until the liquid has been absorbed.
9. Invert the plate and incubate overnight at 37°C.

11. SCREENING OF TRANSFORMED COLONIES

Buffers and Reagents

Transformed colonies, X-gal (20mg/ml); IPTG (200 mg/ml); LB agar plates; ampicillin (100 mg/ml) solutions; Sterile tooth picks.

Protocol

1. Prepare LB-agar plates with 60 μl/ml ampicillin.
2. Add 40 μl of X-gal and 4 μl of IPTG.
3. Spread the solution over the entire surface of the plate. Incubate the plate at 37°C until all fluid disappears.
4. Transfer each individual transformed colony with one tooth pick onto the plate. Upto 100 colonies can be transferred onto a 90 mm plate.
5. Incubate the plate in an inverted position for 12 to 16 hours at 37°C.
6. Observe the blue/white colour of the colonies, the white colonies contain recombinant plasmids while blue colonies contain non-recombinant plasmids.

Subculture of transformed colonies

Inoculate fresh LB culture medium with the transformed colonies (use separate culture tube for individual colonies) following the protocol described in single cell culture (Streak-plate method)

REFERENCES

1. Freifelder, D. 1983, Restriction endonucleases. In Molecular Biology (K. Sergent, ed.) 2nd ed. Jones and Bartlett, Portola Valley, California, pp. 124-129.
2. Grunstein, M. and Hogness, D.S. 1975. Proc Natl, Acad, Sci, USA, 72:3961-3965.

3. Rodriquez, R.L. and Tait, R.C. 1983. Gel electrophoresis. Recombinant DNA Techniques: An Introduction.
4. Addison-Wesley, Reading, Massachusetts, pp67-79.
5. Schleif, R.F. and Wensink, P.C. 1981. Practical Methods in Molecular Biology. Springer-Verlag, Berlin and New York, p 145.
6. Schuler, M.A. and Zielirnski, R.E. Methods in plant Molecular Biology. Academic Press, Inc. San Diego, New York, London and Tokyo P. 19-38.
7. Tang, K. Nakata, Y, Li, H-L, Zang, M, Gao, T., Fujita, A, Sakatsume, O, Ohta, T., and Yakoyame, K. 1994. Nucleic Acids Research, 22: 2857-2858.

28

ISOLATION OF BACTERIOPHAGE LAMBDA DNA

Bacteriophage λ is a double stranded DNA virus with a genome size of approximately 50 kb. In bacteriophage λ particles, the DNA is in the form of linear duplex with single stranded complementary ends 12 nucleotides in length. Inside the host, Bacteriophage lamda DNA circularize through pairing of cohesive ends and exist as a circular molecule. It has two types of growth cycle. (1) Lytic growth cycle and (2) Lysogenic growth cycle. In the present protocol, it is lytic growth cycle of the bacteriophage which has been exploited. Bacteriophage λ has been used very extensively to construct large numbers of different types of vectors for different purposes in molecular biology investigations. Besides it uses in the vectors construction, its DNA digest with different restriction endonucleases, are serving as good source of molecular markers.

1. Culture media and Buffers

NZCYM medium

NZ amine	-	10g
NaCl	-	5g.
Casamino acid	-	1g
Bacto yeast extract	-	5g
$MgSO_4$ $7H_2O$	-	2g
Water	-	1L

ΣM Βυφφερ

NaCl	-	5.8g
$MgSO_4$ $7H_2O$	-	2g
1M Tris-Cl (pH 7.5)	-	50ml
2% gelatin	-	5 ml
Water	-	1L

TE

(10 mM Tris Cl and 1 mM EDTA)

2. Infection of *Escherichia coli* cells with Bacteriophage λ

1. Inoculate 100 ml of LB in a 500 ml flask with a single colony of LE 392. Incubate overnight at 37°C with vigorous shaking.
2. Read OD_{600} of the culture. Calculate the cell concentration assuming that 1 $OD_{600} = 8 \times 10^8$ cells/ml.
3. Withdraw two aliquots, each containing 10^{10} cells. Centrifuge at 8000 rpm for 5 minutes at room temperature. Discard the supernatants.
4. Resuspend each of the bacterial pellets in 3 ml of SM.
5. Add bacteriophage and mix rapidly. The number of bacteriophages used is critical. Approximately 5×10^8 bacteriophage would serve thc purpose.
6. Incubate at 37°C for 20 minutes with intermittent shaking.
7. Add each infected aliquot of 10^{10} cells to 500 ml of NZCYM, prewarmed to 37°C in a 2 liter flask. Incubate at 37°C with vigorous shaking. Concomitant growth of bacteria and bacteriophage should occur, resulting in lysis of the culture after 9-12 hours.
8. Add 10 ml of chloroform to each flask and continue incubating with shaking for a further 30 minutes.

3. Purification of Bacteriophage λ

1. Chill the lysed culture to room temperature and add pancreatic DNase and RNase, both to a final concentration of 1 μg/ml. Incubate for 30 minutes at room temperature to digest the nucleic acids liberated from the lysed bacteria, which might otherwise entrap bacteriophage particles.
2. Add sodium chloride to a final concentration of 1M (29.2 g/500 ml of culture). Dissolve by swirling. Let stand for 1 hour on ice.
3. Remove debris by centrifugation at 11000 g for 10 minutes at 4°C. Pool the supernatants in a clean flask.
4. Add solid polyethylene glycol (PEG 6000) to a final concentration of 10% w/v (i.e. 50g/500 ml of supernatant). Dissolve by slow stirring on a magnetic stirrer at room temperature.
5. Cool in ice water and let stand for at least 1 hour to allow the bacteriophage particles to form a precipitate.
6. Recover the precipitated bacteriophage particles by centrifugation at 11000 g for 10 minutes at 4°C. Discard the supernatant and

stand the centrifuge bottles in a tilted position for 5 minutes to allow the remaining fluid to drain away from the pellet. Remove the fluid with a pipette.

7. Using a wide bore pipette equipped with a rubber bulb, gently resuspend the bacteriophage pellet in SM (8 ml for each 500 ml of supernatant). Wash the walls of centrifuge bottles thoroughly since the bacteriophage precipitate sticks to them, especially if the bottles are old.
8. Add an equal volume of chloroform to the bacteriophage suspension and vortex for 30 seconds. Separate the organic and aqueous phase by centrifugation at 1600 g for 15 minutes at 4°C. Remove the aqueous phase containing the bacteriophage.
9. Collect the bacteriophage particles from aqueous phase by centrifugation at 25,000 rpm for 2 hours at 4°C in a Beckman SW 27 rotor (or its equivalent).
10. Pour off the supernatant. A glassy pellet of bacteriophage should be visible on the bottom of the tube. Add 1-2 ml of SM to each tube and leave it overnight at 4°C, if possible on a rocking platform.
11. The following morning, pipette the solution gently to ensure that all the bacteriophage particles have been resuspended.

4. DNA Extraction of Bacteriophage λ

1. Add pronase to a final concentration of 0.5 mg/ml or add proteinase K to final concentration of 5 μg/ml.
2. Add SDS (stock solution 20%, w/v in water) to a final concentration of 0.5%. Mix by inverting the tubes several times.
3. Incubate for 1 hour at 37°C (pronase) or 65°C (proteinase K).
4. Add an equal volume of phenol equilibrated with 50 mM Tris (pH 8.0). Mix by inverting the tube serval times. Separate the phases by centrifugation at 1600 g for 5 minutes at room temperature. Use a wide bore pipette to transfer the aqueous phase to a clean tube.
5. Extract aqueous phase once with a 50:50 mixture of equilibrated phenol and chloroform.
6. Recover the aqueous phase as described above and extract once with an equal volume of chloroform.
7. Transfer the aqueous phase to a dialysis tube.
8. Dialyze sequentially at 4°C against three changes of a 1000 fold volumes of TE.

PHASE LYSATE PREPARATION

Protocol

1. Inoculate single colony LE 392 (*E. coli* strain of λ host) in 50ml Luria Broth (LB) with 0.2% Maltose (Filter sterilized). Incubate at 37°C in Gyratory shaker at 200 rpm, overnight.
2. Do fresh inoculation with 100 μl of overnight grown culture in 50 ml LB medium with 0.2% Maltose. Incubate at 37°C in Gyratory shaker till cell density reaches an OD_{600} of approximately 0.4 (reaches approx. by 3-3½ hrs.).
3. Centrifuge to collect bacterial pellet at 8000 × g for 5 min at 4°C. Discard the supernatan. Resuspend pellet in 3ml SM buffer by vigorous shaking. Mix 100μl Lamda (λ) phase lysate and incubate with intermittent shaking at 37°C for 20' for phase infection to take place.
4. Transfer the phase-Bacterial mix to 50ml LB with 0.2% Maltose and incubate at 37°C in Gyratory shaker at 200 rpm.

   (Clear lysate appears in about 7-9 hrs.)
5. Add 2.5 ml chloroform to each broth culture, mix and leave at room temp, overnight.
6. Centifuge the culture at 8000 × g, 5 min at 4°C to collect supernatant that contains lysate.

   (Follow further steps given in protocol for phage DNA isolation.)

5. Making DNA from small scale liquid lysate

1. To approximately 50 ml liquid lysate, add 10 μl of 5 mg/ml DNase and 25 μl of 10 mg/ml RNase, incubate 1 hr. at 37°C.
2. Pellet the phage by spinning 1½ hr. at 30,000 rpm in an SW-27 rotor.
3. Pour off supernatant, invert the tubes on an absorbent surface to remove any remaining liquid, resuspend the phage in 200 μl of 0.05 M Tris (pH 8.0) (A small translucent pellet should be visible after the tubes are inverted).
4. Transfer the solution to a microcentrifuge tube and add 200 μl buffered phenol. Vortex 20 min. or shake 20 min in microcentrifuge tube shaker. Spin 2 min in microcentrifuge and save the aqueous (top) layer. Repeat phenol extraction.

   (Phenol denatures the phages capsid and releases the DNA. This denatured capsid protein appears as a thick white precipitates at the phenol/water interface. Vigorous agitation is necessary because the pellet is difficult to resuspend. There should be less white precipitate after the second phenol extraction. If there is still a large amount at the interface, do a third extraction).

5. Add 200 µl chloroform, shake well, and spin in microcentri-fuge briefly. Save the aqueous (top) layer. Repeat this step.
6. Add 20 µl of 3 M sodium acetate, and precipitate DNA with 6 vol. of 100% ethanol, incubate for 30 min on ice.
7. Spin in microcentrifuge for 10 minutes. Remove supernatant.
8. Wash the pellet by adding 1 ml of 70% ethanol, spin for 5 min, and remove the supernatant.
9. Dry pellet under a vacuum and resuspend the DNA in 40 µl of TE buffer.

   (3 µl of the DNA suspension should be used for a restriction digest).

REFERENCES

1. Grossberger, D. 1987. Nucleic Acids Research, 15: 6737 p.p.
2. Schuler, M.A. and Zielinski, RE 1995. Methods in plant Molecular Biology. Academy Press, Inc. New York PP 199-122.

29

SCREENING OF GENOMIC AND cDNA LIBRARIES

Exchange of genetic material rarely occurs between unrelated organisms and even similar species may not exchange chromosomal genes. A certain degree of DNA sequence homology is a prerequisite for genetic recombination *in vivo*. However, recombination between DNA molecules *in vitro* is not subjected to such taxonomic restrictions. Consequently using modern methodology of new DNA technology, genetic material from organisms of quite diverse origins with little DNA sequence homology can now be attached into appropriate vectors and introduced into a host cell for propagation and expression. In this way, a gene of interest can now be isolated from any living source or synthesized *in vitro*. Recombinant DNA is an advancement in high technology which allows manipulation of genomes of higher organisms and dissection of their complex regulatory circuits.

The isolation of a gene begins with the cloning of fragmented DNA of the organism into an appropriate vector. This is called the library which is screened by nucleotide probe or antibody to find the appropriate gene.

SOLUTIONS

1. Denaturing Solution

   0.5 M NaOH, 1.5 M NaCl

2. Neutralizing Solution

   0.5 M Tris-HCl pH 7.4,

   1.5 M NaCl

3. 20 × SSC

3 M NaCl	175.2 g
0.3 M Na_3 citrate	88.4 g
double distilled water	1000ml

4. 100 × Denhardts
   - 2% BSA (Bovine Serum Albumin)
   - 2% Ficoll (Sigma 4000,000)
   - 2% PVP
5. Salmon sperm DNA (10 mg/ml)
6. 500 mM EDTA pH 8.0
7. TE (10 mM Tris-HCl pH 8.0, 1 mM EDTA)
8. Pre-hybridization/Hybridization Solution

20 × SSC	12.5ml
100 × Denhardt solution	2.5 ml
20%SDS	1.5 ml
0.5M EDTA (pH 7.0)	0.25 ml
1M $NaPO_4$	2.5 ml
Dextran sulphate	2.5g

Salmon sperm DNA 50 μg/ml of solution

MEDIA

1. LB Medium (per litre)

Bacto-tryptone	10g
Bacto-yeast extract	5g
NaCl	10g

2. Maltose (20%)

Maltose	20g
H_2O	100 ml

Sterilize by filtration through a 0.2 μ filter cassette.

3. $MgSO_4$ (10 mM)

$MgSO_4.7\ H_2O$	0.24 g
H_2O	100 ml

Autoclave to sterilize.

4. SM (per litre)

NaCl	5.8 g
$MgSO_4.7H_2O$	2g
1M Tris-HCl (pH 7.5)	50 ml
2% gelatin	5ml

Sterilize by autoclaving and store in 50 ml lots.

5. NZY broth

NaCl	5g
$MgSO_4.7H_2O$	2g
Yeast extract	5g
Casein hydrolysate	10g

Dissolve in 900 ml of distilled water. Adjust pH to 7.5 with NaOH and make up volume to 1000ml.

6. Top agar

To the desired NZY broth, add agarose to 0.7%. Melt, dissolve, dispense in 5 ml aliquots and autoclave.

7. NZY media

To NZY broth, add 1.5% agar. Autoclave and pour the plates.

Protocols

(a) Preparation of Plating Bacteria

1. Inoculate a single bacterial colony from a day old plate of the indicator bacteria into 50 ml of LB medium supplemented with 0.2% maltose, 10 mM $MgSO_4$ in a 250ml flask and grow to an OD_{600} = 0.8 Bacteria grown in the presence of maltose adsorb bacteriophage λ more efficiently than if maltose was not present; the sugar induces the maltose operon, which contains the gene (lamb) coding for the λ receptor.
2. Centrifuge the cells at 2000 × g for 10 min at room temperature.
3. Discard the supernatant and resuspend the cell pellet in sterile 0.01M $MgSO_4$ (0.4 × the volume of the original culture). The resuspended cells are diluted to an OD_{600} = 0.5, i.e., 1.5×10^9 cells/ml).
4. Store at 4°C. The bacterial suspension may be used for up to one week. However, the highest plating efficiencies are obtained with fresh cells (0 to 2 days old).

(b) Plating of Library (c DNA library)

1. Prepare 100-fold serial dilutions of bacteriophage stocks in SM. Dispense 10 µl of each dilution to be assayed into Eppendorf tubes.
2. Add 260 µl of plating bacteria to each tube. Mix by shaking or vortexing. Incubate at 36°C for 20 min to allow the bacteriophage particles to adsorb.

3. Add 5 ml of molten (48°, top agarose 0.7% and pour onto a 150 mm LB agar plate. The plates must be dry, otherwise the layer of top agarose peels off with the filter.

4. Incubate at 37°C until the plaques reach a diameter of approximately 1.5 mm and are just beginning to make contact with one another (10-12 hrs). The plate should not show confluent lysis.

5. Chill the plates at 4°C for two to four hours to allow the top agarose to harden.

(c) Transfer of Plaques to Filters

1. Wear gloves and then number membrance filters with a soft pencil.

2. At room temperature, place a dry membrance circle neatly onto the surface of the top agarose of a cold plate with plaques so that it comes into direct contact with the plaques. Be careful not to trap air bubble. The filter should be handled with gloved hands; finger oils prevent wetting of the filter and affect transfer of DNA. Mark the filter in three or more asymmetric locations by stabbing through it and into the agar beneath with an 18-gauge needle attached to a syringe containing waterproof black drawing ink. Once in contact with the top agarose, the filter wets very rapidly and transfer of bacteriophage occurs quickly. Therefore, do not move the filter, once contact with the plate is made. The easiest way of placing the filter on the plate is to hold it by its edges bending it slightly so that the middle of the filter makes contact with the center of the plate. Let wetting action pull the rest of the filter onto the plate. Make certain that the keying marks are asymmetrically placed and that both the filter and the plate are marked. There must be enough ink on the plate to be easily visible when a second filter is in place. Large blotches of ink, however, are undesirable.

3. After 60 seconds, use a blunt ended forceps to peel off the first filter and immerse it, phage side up, in a shallow tray of a denaturing solution (1.5M NaCl, 0.5M NaOH) for 7 min. This process causes denaturation of the protein portions of the phage and releases the DNA which binds to the membrane filter. Transfer the filter into neutralizing solution (1.5 M NaCl, 0.5M Tris HCl, pH 7.0) for 3 min. Rinse the filter in 2 × SSC (diluted from 20 × SSC) for 5 min and place it on whatman paper to dry with DNA side up for 30 min.

4. Place a second, dry filter onto the same plate and mark it with ink at the same locations. After 1-2 minutes, peel the filter off the plate. Denature the DNA and neutralize as described in step 3 above.

5. When all the filters are dry, wrap them between sheets of Whatman 3mm paper. Fix the DNA to the filter by baking for 2 hours at 80°C in an oven. (Overbaking can cause the filters to become brittle). Store in a desiccator until use.

PRIMARY SCREENING

(d) Hybridization with Labelled Probe

Prehybridization and hybridization

Seal each filter into a plastic bag. Cut a corner of the sealed bag and add the prehybridization solution to each bag. For a small filter (82mm) add 5ml of prehybridization solution per bag, while for a 147 mm of filter add 7ml of prehybridization solution to each bag. Squeeze all the air bubbles out and then seal the corner of the bag. Each of these bags containing a single filter should be put into a larger bag and sealed in all four corners. Put this plastic bag containing all the small filters on a platform of G-76 shaker. Put lead weights on all four corners and fill the shaker with 60°C water so that the bag is immersed. Let it shake at a setting of 4.5 for four hrs. After four hours remove bag from water bath. Split the big bag and take out all the small bags. Cut the corner of each small bag and remove all the prehybridization solution. Add 2ml of the hybridization solution per bag. Add ^{32}P-labelled probe. Seal the corners after removing all the bubbles. Put all the small bags into a big bag. Seal all four corners and immerse in a 60°C water bath shaker for 16 hours.

Washing of Filters

1. After 16 hrs of hybridization at 60°C take the filters out of the bags.

2. Wash twice for 30 min at 37°C in 1L of 2 × SSC 0.1% SDS.

3. Wash two times in 1L of 1 × SSC for 30 min at 60°C.

4. Put the filters on 3MM Watman paper to air dry with plaques side up and numbered side down. Let them air dry for one hour.

5. Place these filters on a sheet of 3MM Watman chromato-graphy paper with DNA side up. Make four radioactive markings on 3MM watman paper for later alignment of the radioautographed film. Wrap the 3MM paper with S&S filters in saran Wrap. Load

into cassettes with X-ray film for autoradiography in a dark room. Put at -80°C for overnight exposure and then develop the film.

(e) Interpreting Plaque DNA Hybridization Signals

1. Put 3MM filter paper containing hybridized filters on a light box.
2. Put the developed film on the filter paper and align with the radioactive ink marks.
3. Mark on the film the number of each of the filters and also note the ink marks which match on the hybridized filter, the agar plate and the developed film.
4. Put the film on a light box and the agar plate with plaques on top. Align the marks on the agar plate and the film.
5. With Pasteur pipette, pick the plaques which give positive signals.

(f) Picking Plaques

1. Place 1.0 ml of SM in an Eppendorf tube. Add 1 drop of chloroform.
2. Using a pasteur pipette equipped with a rubber bulb, or a micropipette, stab through the chosen plaque into the hard agar beneath. Apply mild suction so that the plaque, together with the underlying agar, is drawn into the pipette.
3. Wash out the fragments of agar into the 1.0 ml of SM containing a drop of chloroform. Let stand at room temperature for an hour or two to allow the bacteriophage particles to diffuse out of the agar. An average plaque yields 10^6-10^7 infectious phage particles, which can be stored indefinitely at 4°C in SM-Chloroform without loss of viability.

Note: Because bacteriophage λ can diffuse considerable distances through the top agar layer, choose well-separated plaques. For the same reason, it is advisable to pick plaques shortly after the bacterial lawn has grown and the bacteriophage plaques have first appeared.

(g) Secondary Screening

Plate these plaques at a dilution of 10^{-2}, 10^{-4} and 10^{-6} on LB agar plates. Choose agar plates with well dispersed and well isolated plaques. Transfer these plaques to cellulose filters and hybridize against the defined gene. After autoradiography, a well isolated recombinant plaque which hybridizes to the probe is picked from each plate for further work.

(h) Plating Lysate Stocks

To achieve maximum yields, the number of bacteriophages plated should be adjusted so that the outer edges of the expanding plaques just touch after approximately 12 hrs of incubation. An inoculum of 10^{-5} plaque forming units (pfu) is usually sufficient to produce confluent lysis on a 85-mm plate (567 cm^2). By the end of the period of growth, there should be no patches of uninfected bacteria.

1. Mix 10^5 pfu of bacteriophage (or 1/20 of a resuspended plaque) with 0.1 ml of plating bacteria. Incubate at 37°C for 20 min.
2. Add. 3.0 ml of melted top agar/agarose at 47°C, mix, and pour onto a labeled 85mm plate containing 30 ml of hardened bottom LB agar. 2-days old, dried plates give the best results.
3. Invert and incubate the plate for 8-12 hrs. until lysis is confluent.
4. Turn the plate over, add 5 ml of SM, and store the plate at 4°C for several hours with intermittent shaking.
5. With a pasteur pipette, harvest as much as possible of the SM. Add 1 ml of fresh SM and store the plate for 15 min in a tilted position to allow all the fluid to drain into one area. Again remove the SM and combine it with the first harvest. Discard the plate.
6. Add 0.1 ml of chloroform to the pooled SM, vortex briefly, and centrifuge at 4000x g for 10 min at 4°C.
7. Recover the supernatant and add chloroform to 0.3%. The titer of bacteriophage (approximately 10^{10}/ml) usually remains unchanged as long as the stock is stored at 4°C.

Master stocks of important 1 bacteriophages (e.g. vectors), whose genotype has been verified should also be stored at -70°C. Add DMSO to the bacteriophage stock to a final concentration of 7% (v/v). Mix gently. Plunge the container into liquid nitrogen. When the liquid has frozen, transfer the container to a freezer at-70°C for long-term storage.

To recover the bacteriophage, scrape the frozen surface of the liquid with a sterile 18 gauge needle, streak the needle, over the surface of a plate containing indicator bacteria in order to obtain bacteriophage plaques.

REFERENCES

1. Beld, M, Sol, C. Goudsmit, J. and Boom. R. 1996. Nucleic Acids Research, 24: 2618-2619.
2. Cecchini, E., Doming, PJ, Geri, C., Kaiser, K. Sentry, J. and Milner, J. J.- 1993. Nucleic Acids Research, 22: 1322-1323.
3. Gussow, D. and Clackson, T. 1989. Nucleic Acids Research, 17: 4000 pp.
4. Hodge, R., Paul, W., Draper, J. and Scott. R. 1992. The Plant Journal, 2: 257-260.
5. Sambrook, J., Fritsch, E.F. and Maniatis, T. 1989. Molecular Clonning : A Laboratory Manual, USA, Cold Spring Harbor Laboratory Press.

30

ISOLATION OF PLASMID DNA (ALKALINE LYSIS METHOD, MINIPREP)

Buffers and Reagents

Equilibrated phenol [P] (pH 8.0); Chloroform: Isoamyl alcohol (CI) (24:1); 100% Ethanol; 70% Ethanol; solution 1 [50 mM glucose, 25 mM Tris-Cl (pH 8.0), 10 mM EDTA (pH 8.0)]; Solution II (0.2 N NaOH, freshly diluted from 10 N stock and 1% SDS); Solution III (5M potassium-acetate-60 ml, glacial acetic acid-11.5 ml, H_2O-28.5ml); TE buffer (pH 8.0).

Protocol

1. Transfer a single bacterial colony to 2ml of LB medium containing ampicillin (60 μg/ml) in a loosely capped 15 ml tube. Incubate the culture overnight at 37°C with vigorous shaking (300 cycles/min in an incubator shaker).
2. Pour 1.5 ml of the culture into a 2 ml EP tube. Centrifuge at 12000 rpm for 30 sec at 4°C (store the remainder of the culture at 4°C).
3. Remove the medium by aspiration, leaving the bacterial pellet as dry as possible.
4. Resuspend the bacterial pellet in 100 μl of ice-cold solution I by vigorous vortexing.
5. Add 200 μl of freshly prepared solution II. Close the tube tightly and mix the contents by inverting the tube rapidly five times. Make sure that the entire surface of the tube comes in contact with solution II. DO NOT VORTEX. Store the tube on ice.
6. Add 150 μl of ice cold solution III. Close the tube and vortex it gently in an inverted position for 10 sec. Store the tube on ice for 3-5 min.
7. Centrifuge at 12000 rpm for 5 min at 4°C in a microfuge, transfer the supernatant to a fresh tube.

8. Add an equal volume of PCI. Mix by vortexing. After centrifugation at 12000 rpm for 2 min at 4°C in a microfuge, transfer the supernatant to a fresh tube.
9. Precipitate the double stranded DNA with 2 volumes of ethanol (100%) at room temperature. Mix by vortexing. Allow the mixture to stand for 2 min at room temperature.
10. Centrifuge at 12000 rpm for 5 min at 4°C in a microfuge.
11. Remove the supernatant by gentle aspiration. Leave the tubes in an inverted position on a paper towel to allow all the fluid to drain away. Remove any drop of fluid adhering to the walls of the tube.
12. Rinse the pellet with 1ml of chilled 70% ethanol at 4°C. Remove the supernatant and allow the pellet of nucleic acid to dry in the air for 10 min.
13. Redissolve the plasmid DNA in 50 μl of TE buffer (pH 8.0)

PLASMID DNA ISOLATION (MID SCALE)

Solutions

1. | | |
   |---|---|
   | 20% Glucose | 2.25 ml |
   | 0.5 M EDTA, pH 8.0 | 1.00 ml |
   | 1M Tris, pH 8.0 | 1.25 ml |
   | St. H_2O | 45.50 ml |

   Store at 4.C.

   Add 2.0 mg/ml lysozyme immediately before use.
2. | | |
   |---|---|
   | 10N NaOH | 0.4 ml |
   | 20% SDS | 1.0 ml |
   | St. H_2O | 18.6 ml |

   Don't autoclave. Shelf life 1 year at RT.
3. 3M Sodium Acetate pH 4.8 dissolve 40.81 g of sodium acetate.

   $3H_2O$ in 60 ml of H_2O. Adjust pH to 4.8.
4. | | |
   |---|---|
   | 1M Tris pH 8.0 | 2.5 ml |
   | 3M NaOAC pH 4.8 | 1.65 ml |
   | St. H_2O | 45.85 ml |
5. RNase A

   Dissolve pancreatic RNase A at a concentration 10 mg/ml in 10 mM Tris pH 7.9.15 mM NaCl. Heat to 100°C for 15 min and allow to cool slowly to room temperature. Dispense into aliquots and store at-20°C.

6. $T_{10}E_1$ pH 8.0

Tris base	0.1211 g
EDTA	0.0372 g

Adjust pH to 8.0 with HCl. Adjust volume to 100 ml, autoclave and use.

Protocol

1. Grow over night culture of the plasmid containing strain in 10 ml L-Broth with appropriate antibiotic (s).
2. Spin for 5 min in Eppendorf, centrifuge at 10000 × g, pour off suspension, (longer spin makes pellet hard to suspend).
3. Suspend pellet thoroughly in 100 μl of Soln. 1 (add lysozyme 5 mg/1.0 ml Soln. 1) by vortexing immediately after adding Soln. 1. Keep on ice for 5 min.
4. Add 200 μl freshly made Soln. 2 at RT. Mix gently by inverting tubes or vortex. Keep on ice for 5 min. or until SDS distinctly precipitates. Lysate clear at first but gets cloudy as SDS precipitates. (This alkaline pH denatures linear host DNA but hot circular forms). Sample should be viscous and stringy when cap opened at completion of this step.
5. Add 150 μl of Soln. 3 at RT and mix gently by inverting tubes several times. Keep on ice for 5 min. (This high salt soln. 3 mM NaoAC neutralizes the pH and ppts. proteins with SDS, host DNA and RNA).
6. Spin 3 min in Eppendorf centrifuge in cold room or for 5 min at 14K at 4°C in Sorvall. Without disturbing the pellet, carefully remove the supernatant. Put into another Eppendorf tube. Discard the pellet.
7. Add 0.9-1.0 ml cold 100% EtOH. Gently mix and freeze at -70°C for 15 min.
8. Spin in Eppendorf centrifuge for 2-3 min. Remove and discard supernatant.
9. Add 100 μl of Soln. 4 to the pellet and gently vortex till pellet is dissolved.
10. Add 200 μl cold 100% EtOH and freeze at -70°C for 15 min again.
11. Spin in microfuge for 2-3 min and remove the supernatant. Wash pellet with 70% ethanol to remove salts. Dry final pellet for 5 min in vacuum.

12. Suspend pellet in 20 μl of T_{10} E_1 pH 8.0, 0.2 μl of this should be enough to show up on gels.

ELUTION/EXTRACTION OF DNA FROM LOW MELTING TEMPERATURE AGAROSE GEL

Solutions

1. 50 × TAE

Tris	242 g
Glacial acetic acid	57.1 ml
0.5M EDTA (pH8.0)	100mL

Make up the volume to 1000 ml. Autoclave and use.

2. 5M NaCl

NaCl	29.25 g

Water 100 ml. autoclave and use.

Protocol

1. Prepare an agarose gel (0.7%) using low melting point agarose. To 0.42 g low melting point agarose add 50 ml water. Heat until dissolved. Then add 1.2 ml 50 × TAE buffer. Make up to 60 ml final volume with water. Add 1μl of ethidium bromide stock solution (10mg/ml).
2. Cool the gel to 37°C before pouring. This will prevent splitting.
3. Place the gel in the tank while the comb and the gel surround are still in position. Cover the gel with buffer prior to removing the comb and the surround.
4. Prior to loading the samples, the gel should be pre-run for a period of 10 minutes. During this time the voltage should be slowly increased until the required setting is obtained. (This prevents splitting of the gel). Run the gel at up to 7.5 U/cm.
5. To remove DNA view the agarose gel under U.V. light. Care is needed when removing the gel from the plate as the L.M.P. agarose does not have the strength of the ordinary agarose. Remove the required fragments with a sterile scalpel.
6. The agarose is then melted by heating it to 65°C for 5 min.
7. For 500μl gel solution add 50μl of 5M NaCl.
8. Vortex and heat at 65°C for 5 min.
9. An equal volume of phenol is added, the contents vortexed and then centrifuged for 5 minutes. The upper aqueous layer is removed and re-extracted with phenol once.

10. Two volumes of ether is added, mixed by inverting the tube and remove the ether layer.
11. Heat at 65°C for 2 minutes.
12. Add 0.1 volume of 3 M sodium acetate and then 3 volumes of ethanol are added. This is then left at -20°C for overnight and the precipitate collected by centrifugation (10 min.)
13. The pellet is then washed with 70% ethanol, dired under vacuum for 10 min and re-suspended in minimum volume of sterile water or TE buffer.

ELECTROELUTION

1. Restrict the DNA with the appropriate restriction enzyme and electrophoreses restricted DNA in an appropriate agarose gel.
2. View the gel under UV light and cut out the desired band with a sterile scalpel or razor blade. Trim off the excess gel.
3. Take a 4 cm length of narrow dialysis tubing (0.8 cm diam.) washed thoroughly and treated by boiling in 10 mM EDTA for 10-15 min. Secure one end of the tubing by folding it and clipping with a Mediclip (alternatively tie two or three knots in it).
4. Insert the gel slice into the tubing and add not more than 0.5 ml of electrophoresis buffer per slice of gel. Exclude all air from the dialysis tubing and close off the other end with another Mediclip or knots.
5. Add 50 ml of electrophoresis buffer containing 5 µl EtBr into a mini gel tank and submerge the gel in the tubing at right angles to the direction of the current.
6. Electrophorese at 60 mA for about 20 min or until all the DNA has left the gel slice and adhered in a fine sharp fluorescent line (under UV light) to the side of the dialysis tubing.
7. Reverse the polarity and electrophoresis for a further 20 sec to detach the DNA from the wall of the tubing. Open one end of the dialysis bag and remove the DNA solution to a sterile COREX tube or Epperendorf tube. There should be no fluorescence (DNA) left in the gel slice.
8. Add an equal volume of phenol/chloroform (50.50), equilibrated with TE buffer, vortex and then centrifuge for 1 min to separate the phases. Remove the upper aqueous phase. Back extraction of the phenol phase can be performed if isolating small amounts of DNA. Pool the aqueous phase.

9. Extract the aqueous phases twice with an equal volume of a mixture of chloroform-isoamyl alcohol (24:1). Remove the upper aqueous layer each time.

   IMPORTANT:- If the aqueous phase is more than 0.2 ml, extract with several volumes of butanol until the volume of the aqueous layer has been adequately reduced (i.e. until the desired DNA concentration has been achieved).

10. Precipitate the DNA in the usual way (viz. add one tenth volume of 3 sodium or potassium acetate, pH 5.5 mix thoroughly and add 2 to 3 vol of absolute ethanol and keep at -20°C for 30-40 min. Centrifuge, wash the pellet with 70% ethanol, dry and redissolve in sterile distilled water or TE buffer.

REFERENCES

1. Grossberger, D.1987 Nucleic Acids Research, 15:6737.
2. Sambrook, J;Fritsch, E.F. and Maniatis, T. 1989, Molecular Cloning: A Laboratory Manual. Cold Spring Harbor Laboratory Press.

31

PREPARATION OF PROBE

The oligonucleotide to be used as a probe should be purified by gel electrophoresis. The band can be located after staining with ethidium bromide and electroeluted. The probe DNA is precipitated with ethanol and dissolved in 20 μl of TE buffer. Half a microliter of this can be again analysed by a 1% agarose gel electrophoresis together with the dilutions of known amounts of plasmid DNA. By comparing the intensity of DNA bands under UV after ethidium bromide staining, the approximate concentration of probe fragment can be assessed.

1. NICK TRANSLATION

1. DNase 1 (1 mg/ml in 10 mM HCl), aliquoted and stored at -20°C.
2. DNase 1 activation buffer

   10 mM Tris, pH 7.6
3. 10 × NT Buffer

   0.5M Tris, pH 7.8

   5mM $MgCl_2$

   100 mM Beta mercaptoethanol

   500 μg/ml Bovine Serum Albumin
4. α-^{32}P dCTP (>400 Ci/mMole)
5. DNA polymerase 1.
6. 1 mM dATP in water, pH 8.0

   1 mM dGTP

   1 mM TTP

Protocol

To activate DNAse 1, take 2 μl DNAse 1, keep on ice for 2h. 18 μl Activation buffer.

Immediately before use dilute DNAse 1 in a ratio of 1:100 in water (5 μl in 495 μl water) and then 1:100 again (final dilution 1:10,000)

Prepare Nick translation reaction mix

12 μl α-^{32}P CTP (2 × 10^8 CPM)
2 μl DNA fragment (0.1 μg DNA)
2 μl 10 × NT buffer-dCTP
2 μl dATP, dGTP, TTP mix (18 μM each)
1 μl DNase-40Pg/
1 μl DNA Pol 1-1 unit/μl
Incubate at 15°C for one hour and stop by adding 1 μl EDTA (0.5M) and 80μl of H_2O.

Purify the labelled fragment by passing through a Sephadex G-50 column equilibrated with TE pH 8.0. Follow the column by monitoring for radioactivity. Collect radioactive fractions (0.5ml). If necessary concentrate by ethanol precipitation or freeze drying. Use 10^2 to 10^6 cpm per filter for hybridization.

2. RANDOM PRIMED LABELLING OF PROBE DNA

α-^{32}P Random Primed DNA Labelling of probes

Buffers & Solutions

Random Prime Labelling Kit : Klenow Fragment (34/μl): Random hexanucleotide primer (100 ng/μl); l0x labelling buffer [900 mM HEPES (pH 6.6) & 100 mM Mg Cl_2]; dNTPs (10 mM/25μl of each dNTP); Sterile water; dithiothreitol (DTT) (20 mM/50μl); (α^{32}P) dATP (10 μ Ci/μl); Probe DNA.

Protocol

1. Denature 25 ng probe DNA (in a volume of at least 10 μl) by heating for 10 min at 95°C and then chill on ice.
2. Add 2.5 μl of 10× labelling buffer.
3. Add 1 μl of random hexanucleotide primer.
4. Add 2.5 μl of DTT solution.
5. Add 2 μl of dNTP mix (dCTP, dGTP and dTTP, each at concentration of 2.5mM).
6. Add 3 μl of (α-^{32}P) dATP.
7. Add 3 μl autoclaved ddH_2O.
8. Add 1 μl of klenow fragment and mix the reaction gently.

9. Centrifuge briefly to settle the contents in the bottom of EP tube.
10. Incubate at room temperature at least for 2 h and stop reaction by adding 1μl of 0.5 M EDTA (pH8.0)

3. END LABELLING OF OLIGONUCLEOTIDE PROBE DNA

Buffers & Solutions

Solution 1: 10 pmoles oligonucleotide (10 pmoles/μl);
Solution 2: Double distilled water;
Solution 3: 10× kinase buffer;
Solution 4: (α^{32}P) ATP (10 μCi/μl);
Solution 5: T4 Polynucleotide kinase;
Solution 6: 0.5 M EDTA (pH 8.0)

Protocol

Add to a 1.5 ml EP tube, containing 1 μl of probe DNA (10 pmoles), following in the given order:
3 μl Solution 2
1 μl Solution 3
5 μl Solution 4
0.5-2 units Solution 5
Mix carefully and incubate at 37°C for 1 hour. Stop the reaction by adding 1 μl of Solution 6. Leave the mixture on ice until the next step.

4. IN-GEL HYBRIDIZATION

Buffers & Solutions

Solution 1 : 6× SSC

Solution 2: Hybridization solution (10 ml): 2.5 ml 20 × SSPE buffer; 0.5 ml 100 × Denhardt's solution; 0.1 10% SDS; 100 μg *E. coli.* DNA.

Protocol

1. Equilibrate gel in Solution 1 for 5 min.
2. Prepare Solution 2 and prewarm to hybridization temperature (see below for hybridization temperature). Add radiolabelled oligonucleotide probe.
3. Wind the dried agarose gel around a glass pipette and transfer it into screw capped glass hybridization bottle (filled with 6× SSC) by unrolling the gel into the wall of the hybridization bottle.

4. Carefully decant the fluid completely.
5. Add a volume of solution 2 (0.1ml/cm^2 gel) containing radiolabelled oligonucleotides equivalent to 10^6 cpm/ml hybridization solution.
6. Close hybridization bottle carefully with screw cap.
7. Hybridize 2-3h (preferably overnight) at 35°C for $(GATA)_4$ and 45°C for $(CAC)_5$ probe.

Note : Hybridization temperature is kept as Tm-5°C. Tm (melting temperature) may be calculated according to Thein & Wallace rule : 4°C for every C or G, 2°C for every A or T.

5. PREPARATION OF DOT BLOTS

Buffers & Solutions

DNA samples (concentration 1 μg/μl); 20× SSC; Denaturation solution (0.5M NaOH + 1.5M NaCl); Neutralization solution [0.5M Tris-Cl (pH7.2) + 1.5M NaCl + 0.001M EDTA]; Hybond-N^+ membrane; Whatman 3MM filter paper; Food wrap.

Protocol

1. Take 1 μl of DNA samples (1μg) in EP tubes.
2. Heat the DNA samples at 95°C for 10 min, then chill on ice.
3. Add equal volume of 20× SSC.
4. Cut a piece of the Hybond-N^+ membrane to the appropriate size and mark a grid of circles using a light pencil.
5. Wet the membrane with 10× SSC for 2 min, air dry it on a piece of 3MM whatman filter paper.
6. Spot DNA samples (2μl) onto the circles marked on the membrane, allow to dry.
7. Wet the membrance in denaturation solution for 5 min.
8. Transfer the membrane to a Whatman 3MM filter paper soaked in neutralization solution for 1 min and allow the membrane to air dry.
9. Wrap the membrane in food wrap and place the DNA side down on a UV transilluminator for 5 min.

6. DNA ESTIMATION BY SPECTROPHOTOMETER

Buffers & Solutions

DNA samples; TE buffer; Sterile double distilled H_2O; 1 ml Quartz cuvettes (two).

Protocol

A. Estimation

DNA yield can be checked by comparison with the known standards by gel electrophoresis or by mixing 1 μl of your sample with a small amount of ethidium dye standard DNA concentrations. It can also be checked spectrophotometrically from the absorbance data of your DNA sample at 260 nm ($1OD_{260}$ = 50μg/ml).

B. Purity Test

Purity of the DNA sample is calculated from

$$\frac{OD\ 260}{OD\ 280}$$

For pure DNA this ratio remains around 1.8.

C. Optical density (OD)

Nucleic acid optical density (OD) is measured in a UV light spectrophotometer. In most instruments a beam of UV is split into two parallel beams of equal intensity which pass through parallel-sided quartz glass sample cells (cuvettes) of fixed light-path length. One cuvette is referred to as the sample cell and the other as the reference cell, the former containing the sample of nucleic acid to be measured and the latter, a sample of the solvent in which it is dissolved.

Light is absorbed to a greater extent in the sample cell than in the reference cell and the absorbance difference between sample and reference is proportional to the concentration of the nucleic acid (or nucleotides) dissolved in the sample.

When using a standard 1 cm light path cell, the OD value refers to the concentration of nucleic acid in 1 cm^3 of the sample. An absorbance value of 1 is referred OD_{260} to as 1.00 OD. One OD absorbance of DNA in a UV beam with transmission of 260 nm (OD) represents a sample concentration of 50 μg double stranded DNA/ml.

D. OD Measurement

1. Make an accurate 1:1000 dilution of the DNA in distilled H_2O (or any other solvent used for DNA) and mix well.

2. Fill the reference cell three quarters with water or the solvent used (1 ml), making sure not to leave bubbles which will defract the light beam and introduce errors, similarly fill the sample cell with your diluted DNA (1 ml).
3. Adjust the spectrophotometer sequentially to measure the OD at 260 and 280 nm.

REFERENCES

1. Evans, M.R. and Read, C.A. 1992. Nature, 358 : 520-521.
2. Feinberg, A.P. and Vogelstein, B. 1984. Analytical Biochemistry, 137 : 266-267.

32

Isolation of High Molecular Weight Plant Genomic DNA and Mitochondrial (mt) DNA

PLANT DNA ISOLATION

Extraction buffer

1M tris-HCl pH 7.5

2M NaCl

0.5 M EDTA

10% SDS

TE buffer

10mM tris-HCl pH8.0

1 mM EDTA

Protocol

1. Grind 0.5g plant material on ice with pestle and mortar. Transfer to an Eppendorf tube and add 0.5ml extraction buffer.
2. Leave at room temperature from 10 minutes to 2 hours.
3. Add equal volume phenol/chlorophorm/isoamyl alcohol (24:24; 1) and invert carefully approximately 10 times.
4. Centrifuge 15 minutes at 8000 rpm and transfer the upper phase to a new Eppendorf tube.
5. Add equal volume phenol/chlorophorm/isoamyl alcohol (24:24:1) again. Turn the tubes 3-4 times and centrifuge at 8000 rpm for 10 minutes.
6. Add in a new Eppendorf tube to the upper phase 1/10 volume 2M sodium acetate pH 4.8 (cold) and equal volume isopropanol and mix carefully.
7. Take DNA with Pasteur pippet, wash 2 times with 500 µl 70% cold ethanol.

8. Centrifuge to collect DNA at the bottom of the tubes. Dry the pellet and dissolve in 50μl TE buffer.

ISOLATION OF SUNFLOWER MITOCHONDRIAL DNA (mt DNA)

Plant material: Etiolated 20 days old seedlings of different sunflower lines (near isogenic, male sterile/male fertile lines and fertility restorer lines).

Protocols

1. Homogenise 100g of plant material in 400 ml ice-cold extraction buffer (0.44 M Mannitol, 50 mM Tris HCl pH 8.0, 3 mM Na_2 EDTA; 0.1% BSA, 0.1% Polyclar AT, 0.15% 2-Mercaptoethanol) for 2× 10 seconds with a blender at low speed.
2. Pass through four layers of cheese cloth and two layers of mircloth and centrifuge at 3000 × g at 4°C for 10 min.
3. Discard the pallet and take the supernatant and centrifuge it at 15000 × g at 4°C for 15 min.
4. Discard supernatant and resuspend the pellets in 0.1 ml/g buffer (0.3 M Sucrose, 5 mM Tris HCl pH 7.5) and digest with 50g/ml DNAse for 60 min at room temperature in the presence of 10mM $MgCl_2$ and 0.05% BSA.
5. After DNAse treatment, put a layer of 20ml Shelf buffer (0.6 M Sucrose, 10mM Tris HCl, 2mM Na_2EDTA pH 7.2) under the digestion mixture and centrifuge it at 10000× g for 20 min.
6. Discard supernantant and wash the pellets with 2 × 10 ml Shelf buffer and collect the purified mitochondria.
7. Mitochondria are lysed in 0.1 ml/g lysis buffer (500mM Tris HCl pH 8.0, 10 mM Na_2EDTA, 1% SDS and 20 g/ml Proteinase K) for 30 min at 65°C. CsCl density gradient certifugation of the crude lysate with ethidium bromide results in two distinct DNA bands: The lower much weaker band is made up of traces of nuclear DNA. Recover mt DNA from CsCl gradient by piercing from the side and transfer the mtDNA to another tube.
8. Add an equal amount of isopropanol saturated with CsCl saturated TE buffer to remove EtBr. Mix gently and after phase separation, recover upper phase into new tube. Repeat this extraction twice. Complete removal of EtBr can be monitored by a brief inspection under UV illumination.
9. Add 2-3 vol of double distilled water and precipitate DNA with atleast 2 vol of absolute ethanol. After incubation for at least 1 h at

-20°C, pellet the mt DNA by cetrifugation at 12000 × g for 5 min in an Eppendorf cetrifuge, remove supernatant and air dry the mt DNA pellets.

10. Dissolve the mt DNA in 20-100 of TE for further analysis and photometrically determine the concentration of the mt DNA by absorption measurements at 260 and 280 nm. The ratio 260':280 should be 2.1 for good quality mtDNA.

REFERENCES

1. Punia, M.S. 1990. Ph.D. Thesis. Institute of Genetics (Bulgarian Academy of Sciences), Sofia (Bulgaria)
2. Spassova, M.Terachi, T., Usunova, M., Petrov, P., Christov, M., Ivanov, P., Atanassov, A. and Dudov, K. 1992. Biotechnology & Biotechnological Equipments, 6:24-28.

33

RESTRICTION FRAGMENT LENGTH POLYMORPHISM (RFLP)

In recent years gene cloning technology has led to fundamental advances in many areas of biology. One of the most exciting of this is the use of cloned fragment of chromosomal DNA as genetic markers, usually termed as "RFLP mapping" with RFLP denoting Restriction Fragment Length Polymorphism. This technique, depends on natural variation in DNA base sequence where DNA is digested with a restriction enzyme. Homologous restriction fragments of DNA which differ in size or length can be used as genetic markers to follow chromosome segments through genetic crosses. One of the ways to reach this variation is by restriction enzymes. A large piece of DNA is reduced to a series of smaller fragments of defined size by digestion with restriction enzymes. The number of fragments produced and the size of each fragment will reflect the distribution of restriction sites in the DNA. The fragments produced will thus be specific for each target DNA/restriction enzyme combination and can be used as a "finger print" specific for a given target DNA or for the organism containing that DNA.

Advantages of RFLP markers.

1. RFLP markers are co-dominant.
2. Allelic variation is much greater than that for morphological markers.
3. These markers are phenotype neutral.
4. These markers are free of epistatic interactions.

Scope of RFLP studies

1. Abundant natural variation in DNA sequences provides unlimited number of RFLPs. This is helpful in construction of saturated linkage maps.

2. RFLPs' closely linked to qualitative and quantitative traits, can be used for tagging purposes.
3. RFLPs can be used in directed as well as compressed (in time and space) breeding activities.
4. RFLPs can be used in strain and varietal identification.
5. RFLPs can be employed in germ plasm cataloguing.

I. ISOLATION AND PURIFICATION OF HIGH MOLECULAR WEIGHT PLANT DNA

For isolation of plant DNA use protocols discribed in the previous Chapter. However, an another modified method for DNA isolation is given below.

A. Requirements

DNA Extraction buffer

100 mM Tris-Cl (pH 8.0); 20 mM EDTA (pH 8.0); 1.4 M NaCl; 2.0% (W/V) CTAB: 0.2% 2-Mercaptoethanol

Chloroform-Isoamyloalocohol (24:1)

TE Buffer : 10 mM Tris (pH 8.0); 1 mM EDTA

Ethidium Bromide 10 mg/ml

B. Protocol

NA Isolation

1. Grind 5g of frozen leaf material in liquid nitrogen to fine powder.
2. Transfer the powder to a 50 ml polypropylene tube containing 25 ml of pre-warmed DNA extraction buffer. Use spatula to disperse the material completely.
3. Incubate at 60°C for one hour with occasional mixing by gently swirling.
4. Add 15 ml chloroform-isoamyl alcohol and mix by inversion to emulsify.
5. Spin at 15000 rpm for 10 minutes at room temperature.
6. Take out the aqueous phase to another tube, add 2/3rd volume of isopropanol and mix by quick gentle inversion and centrifuge for 3min at 10000 × g.
7. Loop out DNA by using a bent Pasteur pipette.
8. Wash in 70% ethanol, 10 mM ammonium acetate for 20 minutes.
9. Dry the pellet and dissolve in lμl of TE buffer.

Purification of DNA by ultracentrifugation

1. Add 4.5 g of CsCl and 200 μl ethidium bromide.
2. Pellet the precipitate, if any, and transfer the supernatant using syringe (5ml) and needle (18 gauge) to polyallomer quick seal centrifuge tubes.
3. heat-seal the tube after proper balancing.
4. Spin the tube at 60,000 rpm in Ti 65 rotor for 18 hrs at 20°C.
5. Take out the DNA band with the help of syringe (2ml) and needle (18 gauge) and transfer to Eppendorf tube.
6. Add equal volume of butanol saturated with TE to remove Ethidium bromide.
7. Mix by gentle inversion and spin for 2 min.
8. Carefully remove the upper phase and discard.
9. Repeat steps 6-8 till the lower phase become colourless.
10. Transfer the lower phase to a 30 ml corex tube.
11. Add two volumes of sterile distilled water and mix.
12. Add two volumes of cold ethanol and mix by quick gentle inversion.
13. Spool out DNA, wash for 10 min in 70% ethanol and 10 mM ammonium acetate.
14. Dry, dissolve DNA in 500 μl of TE and estimate concentration spectrophotometrically.
15. Store DNA at-20°C.

(II) RESTRICTION ENZYME DIGESTION AND AGAROSE GEL ELECTROPHORESIS OF PLANT DNA.

The choice of which enzyme is to use depends on the cost and the ability of enzymes to detect polymorphisms. Commonly for RFLPs enzymes that have 6 bp recognition sites are used (*EcoRl, EcoRV, HindIII*), etc. because they cut DNA not so often as *Mspl, Taql.* 4bp recognition sites which generate fragments between 300bp and 20kb.

The amount of enzyme used should be in range 3 to 10 units per microgram of DNA and the reaction should be proceed for a few hours to overnight.

The unit definition is the amount of enzyme needed to digest 1 microgram of standard purified DNA (usually lambda phage) in 1 hour.

Often the excess enzyme is needed to be used for plant DNA to ensure complete digestion.

The majority of enzymes have optimum reaction at 37°C but some as *Taql.* which is isolated from termophilic organism have higher optimum 65°C temperature.

Enzymes are stored at -20°C and should always be kept on ice during use. The amount of DNA to be digest should reflect the size of genome under investigation. Large genome/grasses/may require 10 microgramme per gel line in order to obtain significant radioactive signal.

Most gels accommodate between 30 and 50 µl sample size. If DNA too dilute to digest after a digestion the sample is need to be concentrated (by ethanol precipitation).

A. Requirements

10× TBE Bufer

0.89 M Tris-base; 0.89 M Boric-acid; 0.02 M EDTA Ethidium Bromide (10 mg/ml)

10 × Gel loading Buffer.

0.25% Bromophenol Blue; 0.25% Xylene cyanol; 25% Ficoll (type 400) in water;

5M Ammonium Acetate

0.5.M EDTA, pH 8.0

B. Protocol

RESTRICTION

1. Decide the total volume of reaction mixture. Let it be 25 µl.
2. Calculate the volume of DNA stock required for 10 µg DNA to be used.
3. Calculate the volume of restriction enzyme which is to be added. (Generally at the rate of 4 units/µg of DNA).
4. Add the components of the reaction mixture to an.Eppendorf tube in the following order.

   a) Water, b) Buffer (2.5 µl), c) DNA, d) Restriction Enzyme.
5. Mix by gentle tapping and incubate at 37°C overnight.
6. Add 1 µl of 0.5 M EDTA to stop the reaction.
7. Store at 4°C.

ELECTROPHORESIS

1. Place the gel casting plate over a levelled surface. Seal the sides using cellophane tape.
2. Choose an appropriate comb and fix into position towards one end of the plate. Adjust the teeth so that they are 2 mm above the surface of the plate.
3. Add 0.8 g of agarose to 100 ml TBE Buffer (0.5×).
4. Heat in a boiling water bath until agarose dissolves.
5. Cool to 50°C and pour into the casting plates.
6. Allow it to cool and set for 30 min.
7. Pour some TBE (0.5×) buffer onto the gel and remove the comb carefully.
8. Take out the cellophane tapes and transfer the gel plate to the electrophoresis unit.
9. Add enough running buffer (0.5× TBE) so as to cover the gel completely.
10. Mix the restricted DNA samples with 1/10 volume of gel loading buffer and load on to the gel. Include; λ HindIII digest as molecular weight marker.
11. Apply 30 V and allow run to continue for 12-14 hrs.
12. Remove the gel and stain with ethidium bromide (EtBr) (1mg/ml water) for 15-20 minutes.
13. View the gel in dark under UV light.
14. Mark the position of well and molecular weight markers on a paper.
15. Photograph the gel and then proceed for southern blotting.

(III) SOUTHERN BLOTTING OF DNA

Buffers and solutions

TBE buffer

Concentrated stock solution 5× (perL)	Working concentration 0.5 ×
54 g Tris base	0.045 M Tris-borate
27.5 g Boric acid	
20 ml 0.5 M EDTA, pH 8.0	0.001 M EDTA (pH8.9)

20 × SSC

Na_3 citrate	0.3M
NaCl	3M

Gel Electrophoresis

Agarose gel electrophoresis is conducted in a horizontal configuration, because it (1) provides better support at low agarose concentration, (2) there is less distortion during electrophoresis and (3) the bands of DNA are less distorted.

1. Make 0.8% agarose solution in 1 × TBE buffer. Heat the solution up to approximately 90°C for dissolving the agarose completely.
2. Cool agarose solution to 50-60°C. Add ethidium bromide, mix, pour into gel former, and insert gel comb. Remove any air bubbles.
3. Allow it to solidify and place gel in electrophoresis tank. Fill tank with gel running buffer, sufficient to cover surface of the gel.
4. Apply DNA samples to wells in gel. Apply current to gel and allow to run until bromophenol blue indicates that sample has run at sufficient distance.

Gel processing

1. Slide gel from the gel former into a suitable plastic box or a glass/pyrex dish.
2. Cover the gel with 0.25 N HCl (Depurination solution) and begin agitation using orbital shaker, ensuring that the gel moves freely. Treatment should be stopped immediately after the bromophenol blue changes its color (10-15 minutes).
3. Discard the depuration solution and rinse the gel with distilled water. Cover gel with 1.5 M NaCl, 0.5 M NaOH and begin agitation. Agitate for 30 minutes after the bromophenol blue dye has returned its blue colour.
4. Discard the solution and rinse the gel with distilled water. Repeat the step 3 one more time.
5. Cut the membrane (Gene Screen + or Nybond) to the size of the gel and soak it for 10 minutes in distilled water and for 1 minute in 1.5 M NaCl, 0.5 M NaOH.
6. Place the gel on the glass plate (upside down)
7. Place the membrane on the top of the gel without air bubbles. Cut 5 sheets of 3 MM Whatman filte paper to the size of the gel, soak in the same denaturation solution and put on the membrane.
8. Place on top of the 3 MM filter paper a stock of about 6 cm paper towels cut to the size of the gel.
9. Allow the transfer to occur at least 2 hours (usually overnight).

Processing the blot

1. Dismantle blotting stack down to membrane, remove membrane and gel together, place membrance side down on a clean piece of 3 MM paper. Peel off the gel and discard. Mark the membrance so that the face carrying the DNA is distinguishable.
2. Place membrane in 6× SSC and rinse for 1 minute with gentle agitation to remove agarose.
3. There are several alternative methods which may be used for DNA fixation-baking or UV treatment are most usually used. Baking for 2 hours at 80°C or UV fixing according to the membrane supplier instructions.
4. Store the membrane under vacuum at room temperature.

   Another method for southern blotting is by vacuum whose protocols are described below. (use either of the two)

Vacuum blotting of DNA from Agarose gel

Buffers and solutions

1. Depurination solution

   0.25 N HCl
2. Depasturation solution

   0.4 N NaOH-0.6M NaCl
3. Neutralisation solution

   1.5 M NaCl-0.5 M Tris-HCl, pH 7.5
4. Transfer solution

   10× SSC (1.5 M NaCl, 0.15 M sodium citrate)
5. 0.4 N NaOH
6. 0.2 M Tris-HCl, pH 7.5-2× SSC

Protocol

1. Incubate gel in Depurination solution for 10 minutes at room temperature.
2. Wash gel in sterile distilled water and incubate in Denaturation solution for 30 minutes at room temperature.
3. Remove gel, wash in distilled water and incubate in Neutralisation solution for 30 minutes at room temperature.
4. Cut Gene screen plus nylon membranes according to the size of the gel and place a mark on its concave side.

5. Carefully place the membrane onto the surface of deionised water and allow membrane to wet by capillary action.
6. Remove the Gene screen plus and lay onto 10× SSC. Let it soak for 15 minutes.
7. Cut a plastic mask, and make an opening on window for the gel such that the gel is slightly bigger than the window.
8. Connect the LKB vacuum blotting unit with the vacuum pump and inline liquid trap.
9. Place the porous support screen with the shiny side up, on the inner rim of the base unit.
10. Place the plastic mask with the window on the support screen.
11. Position the pre-treated nylon membrane carefully within the window. Avoid trapping air bubbles between the membrane and the porous support screen.
12. Slide the gel from the casting tray onto the membrane to fill the window. Avoid trapping air bubbles.
13. Fit the top frame by evenly tightening the four locking nuts.
    Switch on the vacuum pump to mobilise the gel.
14. Mark the well position with a pencil and seal the wells with molten agarose.
15. Pour on the transfer solution to cover the gel and leave for one hour.
16. Remove the transfer solution, turn off the pump and take the gel out.
17. Remove the membrane, immerse in an excess of 0.4 N NaOH for one minute.
18. Take out and immerse the membrane in an excess of 0.2 M Tris-HCl, pH 7.5, 2× SSC.
19. Allow membrane to dry at room temperature.
20. Bake the membrane at 80°C for 2 hrs under vacuum.
21. Store the membrane under vacuum at room temperature.

In vitro LABELLING OF DNA BY NICK TRANSLATION

In vitro labelling of DNA by nick translation is the main technique for producing radioactively labelled DNA. It is based on the ability of *E. coli* DNA polymerase-I to add nucleotide residues to the 3'-hydroxyl terminus that is created when one strand of double stranded DNA molecule is nicked. In addition, the enzyme has a 5' to 3' exonuclease activity that can remove nucleotides from the 5' end of the nick. The elimination of nucleotides from the 5' side and sequential addition on 3' side results in movement of the nick (nick-translation) along the DNA. Thus, by replacing the pre-existing nucleotides with highly radioactive nucleotides, it is possible to prepare α-^{32}P labelled DNA. The radioactive DNA can later be used in hybridization to detect the presence of homologous sequences.

A. Requirements

1. Nick translation kit containing
   (i) Labelled deoxy NTP.
   (ii) Nick translation buffer including cold nucleotides (solution-I)
   (iii) *E. coli* DNA polymerase-I and Deoxyribonuclease (solution-2).
2. Sodium dodecyl sulphate (10%)
3. Sephadex G-50.
4. 10 mM EDTA and 0.2% SDS.

Protocol

1. Set up nick translation reaction as follows in the Eppendorf tube (the volume vary with different nick translation kits).
   (i) Quantity of DNA to be labelled (1 μg) - 10 μg
   (ii) Volume of nucleotide/buffer (sol. 1) - 10 μl
   (iii) Volume of water - 20 μl
   (iv) Volume of (α-^{32}P) dCTP solution (50 μl) - 5 μl
   (v) Volume of enzyme solution (sol. 2) - 50 μl
2. Incubate in a water bath at 15°C for 90 min.
3. Terminate the reaction by adding SDS to the final concentration of 0.5% (i.e. add 2.5 μl of 10% SDS).
4. Heat at 50-60°C for 5 min.
   Typically, optimum incorporation (20-40% incorporation of free label) is achieved within 20-90 min.
5. Separate the nick translated DNA from unincorporated labelled dNTPs by chromatography through a small column of Sephadex G-50.
6. Prepare a Sephadex G-50 column in a 5ml pipette plugged with a sterile glass wool (add 3.0 g of Sephadex G-50 in 250 ml of

column buffer and let it stand overnight at room temperature and pour into the column).

7. Apply the DNA sample to the column. Wash the tube with 100 µl of column buffer and apply to the column.
8. Elute using the column buffer with a 0.5 ml/fraction into different fractions in Eppendorf tubes.
9. Follow the progress of incorporated and unincorporated α-^{32}P/dCTP down the column using a hand-held mini monitor. The leading peak should be collected into Eppendorf tubes as it elutes from the column.
10. Denature the pooled DNA in a boiling water bath for 5 min before using it for blot hybridization.

(v) SOUTHERN HYBRIDIZATION

A. Requirements

1. Prehybridization Buffer

   5× SSC

   0.05 M Sodium Phosphate Buffer

   2.5× Denhart's solution

   2.5 mM EDTA 0.6% SDS

   5.0% Dextran sulphate in sterile distilled water.
2. Washing solutions

   2× SSC-0.1%

   SDS Ix SSC

   0.5× SSC

   0.1× SSC

Protocol

1. Place the baked nylon membrane in hybridization bag.
2. Calculate the volume of the prehybridization buffer according to the size of the membrane. Buffer has to be added at the rate of 70 µl/cm^2 of the membrane.
3. Prewarm prehybridization buffer at 65°C.
4. Calculate the amount of salmon sperm DNA to be added. It is added at the rate of 100 µg/ml of prehybridization solution.
5. Take salmon sperm DNA in an Eppendorf, denature for 10 min in boiling water bath and put in ice immediately.

6. Add the prehybridization solution along with salmon sperm DNA to the hybridization bag.
7. Squeeze out the air bubbles carefully and seal the bag.
8. Place the bag in water bath maintained at 65°C and leave overnight or for minimum eight hours.
9. Prepare probe as described by nick translating the whole plasmid carrying genomic DNA insert.
10. Denature the probe DNA in boiling water for 10 minutes and put in ice immediately.
11. Take out the bag from water bath, cut one corner, and add the probe.
12. Carefully remove air bubbles and seal the bag.
13. Allow hybridization to continue for 20-24 hrs at 65°C.
14. Take out the bag, cut one corner and remove the buffer. Discard buffer from the bag into the bottle meant for the purpose carefully.
15. Prewarm the washing solutions at 65°C. Maintain one shaker at 65°C.
16. Wash the membrane at 65°C for 30 min with each of the solutions. Monitor the count on the filter by mini counter.
17. Wrap the membrane with saran wrap and expose in dark the X-ray films, inside a cassette meant for the purpose.
18. Allow exposure to continue at -70°C for 1 to 7 days depending on count on the filter.
19. Remove the X-ray film in dark and develop the auto-radiogram as described in the following section.

(VI) AUTORADIOGRAPHY

Radioactive nucleic acids and proteins may be detected by autoradiography or by liquid scintillation spectroscopy. Autoradiographic methods, however, are considered superior as they are more sensitive and give higher resolution.

A. Requirements

(i) **X-ray film developer**

Water at 52°C	500 ml
Metol	1g
Sodium sulphite	75g

Hydroquinone	9g
Sodium carbonate (monohydrate)	25g
Potassium bromide	5g

Dissolve in the given order and make up to 1 litre with cold water.

(ii) **Fixer**

Sodium thiosulphate	80g
Potassium metabisulphate	5g

Dissolve in the given order in 250 ml water and filter

(iii) **Stop bath (water containing 2% acetic acid)**

B. Protocol

1. Remove the film after exposing and develop it in developer for 3-5 min in dark.
2. Wash the film in stop bath for 1 min and fix the film in fixer for 5 min in dark.
3. Wash the film under running water for 30 min and dry it at room temperature.
4. Do not reload a moist or cold cassette. Dry before reusing.

C. Precaution

Typically ^{32}P samples are covered with clear plastic wrap (Saran Wrap is recommended). This protects the cassette and intensifying screen from contamination with penetrating radioactivity. The plastic wrap may block a significant amount of the ^{32}S signal. If ^{35}S is used as the radioactive source, do not cover with the plastic wrap unless sample is sufficiently radioactive to penetrate plastic wrap.

Special Precautions for Use of Radioisotopes

1. When working with radioactive material, wear lab coats (these should always be left in the laboratory) and disposable vinyl gloves to minimize the chance of personal contamination. These items will be available in the laboratory.
2. Do not pipet by mouth. Most of the operations will employ very small volumes of liquid that must be dispensed with automatic pipetters. If standard pipets must be used, fill them by suction with a pipet-filling device.
3. Work on a square of absorbent paper backed with polyethylene to prevent contamination of the lab bench. Keep all work items on

this paper. Mark one corner and designate it as a "hot area." A small radioactive trash container should be kept there as well as all items that are suspected to be contaminated (a small, plastic weighing boat works well for this purpose).

4. Put contaminated, disposable items (*e.g.*, pipet tips, microfuge tubes) into the plastic radioactive trash in your radioactive "hot area." At the end of the period, put this trash and any other contaminated disposable items into the trash can marked "Radioactive Waste."
5. Put contaminated glassware (*e.g.* the beaker used for the TCA precipitation assay) in a specially marked dishpan containing Isoclean, a special detergent for decontaminating radioactive labware.
6. After completing work, check your hands and the lab bench for contamination. Use the lab monitor set at the most sensitive setting. It is a good idea to work using the "buddy" system; one lab partner handles the radioactive material; the other lab partner remains "cold" to perform monitoring and handling of non-radioactive solutions. It is also a good idea to wash your hands when you are finished handling radioactive materials, even if no contamination is detected.
7. The organic solvents in liquid scintillation fluid are flammable and toxic when inhaled. Use scintillation fluid in the hood. Cap scintillation vials tightly before leaving the hood area.

REFERENCES

1. Ahn, S.N. Bollich, C.N., Mellung, A.M., and Tanksley, S.D. 1993. Proc-Natl. Acad. Sci. USA. 90: 7980-7984.
2. Anon. 1998. Training Manual on Recent Techniques in Plant Molecular Biology and Biotechnology. NRC on Plant Biotechnology, IARI, New Delhi (India).
3. Devos, K.M., Millan. T. and Gate, M.D. 1993. Theor, Appl. Genet, 85: 784-792.
4. Sanghai Maroof. M.A., Tang, G.P. Biyashev, R.M., Manghan, P.J. and Zhang. I. 1996. Theor. Appl. Genet.
5. Southern, E.M. 1975. J Mol. Biol., 98:503.

34

Isolation of Molecular Markers using Random Amplified Polymorphic DNA (RAPD)

Random Amplified DNA Polymorphism markers (RAPDs) are based on modification of the Polymerase Chain Reaction (PCR). Unlike the conventional PCR, this assay does not require any specific sequence information on the target genome, single oligonucleotides of arbitrary sequence are used to amplify genomic DNA sequences. The number and the size of the fragments generated depend on the nucleotide sequence of the primer used and the source of the template DNA. Thus, a spectrum of PCR products, a genome specific "Fingerprint" is obtained (1,2)

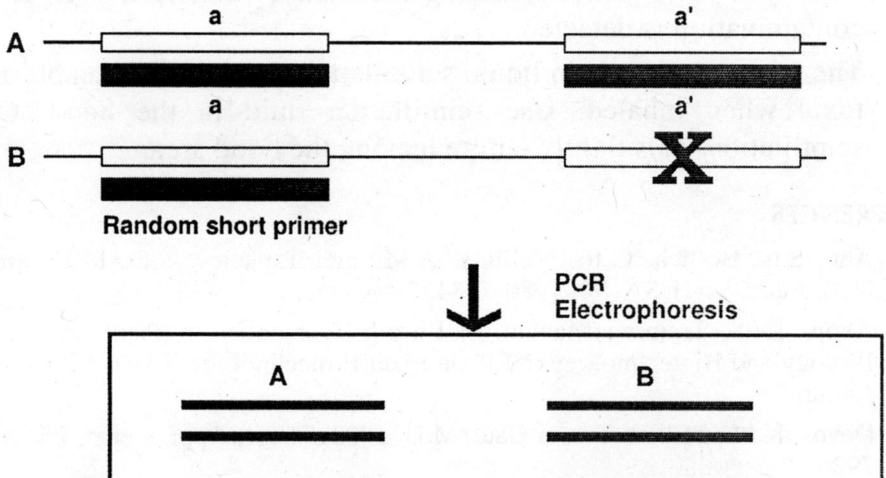

Fig. 34. RAPD marker results from sequence polymorphism between individuals A and B at one of the annealing sites (a′) for a specific RAPD primer. The rest of the RAPD fragments are produced from homo-sequential parts of the genomes A and B.

By comparing the "fingerprints" resulting from the same primer but made on different sources of template DNAs polymorphism can simply be identified as presence or absence of an PCR product on a EtBr stained agarose gel. These polymorphic products are called RAPD markers and are inherited in a Mendelian fashion. The RAPD assay has been used for assessment of variation, cultivar fingerprinting, pathotype identification and genetic mapping.

This protocol will include

1. Use of RAPD assay for fingerprinting the genomes of cultivated tomato (*Lycopcrsicon esculentum*) and its wild relative *L. pennellii*. Because these two species are only distantly related, the RAPD fingerprints are expected to differ markedly.
2. Identification of chromosome 6-specific RAPD markers in tomato (3). By comparing the spectrum of amplified products from *L. esculentum, L. pennellii*, and the *L. esculentum chromosome* 6 substitution line WSL6, which carries chromosome 6 from *L. pennellii*, will identify chromosome 6-specific RAPD markers. Fragments, amplified on DNA from both *L. pennellii* and the chromosome 6 substitution line, but not on DNA from *L. esculentum*, are derived from chromosome 6 of *L. pennellii*. Conversely, amplification products generated on *L. esculentum* DNA, but not on DNA both *L. pennellii* and the chromosome 5 substitution line, should correspond to chromosome 6 of *L. esculentum*.

PROCEDURE

Precautions: Because PCR provides a method to produce many copies of a specific nucleotide sequence from minute quantities of DNA, precautions should be taken to avoid contaminations.

- — autoclave solutions
- — aliquot reagents
- — wear disposable gloves
- — avoid splashes
- — use positive displacement pipettes
- — pre-mix reagents
- — add DNA last
- — choose positive and negative controls carefully

Stock solutions

primers (OPAH13, OPC8 and OPF1)	0.025μg/μl
10× PCR buffer:	
0.5 M KC1	
0.1 M Tris. Cl pH 8.3	
15 mM $MgCl_2$	
0.1% gelatin	
dNTPs	1 mM of each dNTP
Templates-genomic DNAs from	5U/μl
L. esculentum cv. Moneymarker	40ng/μl
L. esculentum chromosome 6 substitution line WSL6	
L. pennellii LA 716	

Amplification reactions

Prepare a mixture of the PCR reagents for each of the primers OPA8, OPC8 and OPFl:

	in each tube	for 12 tubes
10× amplification buffer	4 μl	48 μl
1mM dNTPs	4 μl	48 μl
primers AH13, C8, F1 (from Operon)		
sterile H_2O	17.8 μl	203.6 μl
enzyme Taq DNA polymerase	0.2 μl	2.4 μl

Divide in 0.5 μl sterile microfuge tubes (12 for each primer)

Add:

in tube 1-Moneymarker DNA	10 μl
in tube 2-WSL 6 DNA	10 μl
in tube 3-LA716 DNA	10 μl
in tube 4-sterile water	10 μl

Close the tubes, mix and spin down.

Overlay the reaction mixture with 50 μl of mineral oil (this prevents evaporation of the sample during repeated cycles of heating and cooling).

Transfer the tubes in the PCR machine and carry out amplification at following conditions:

5 min	93°C

50 Cycles:

1 min	93°C
2 min	35°C
2 min	72°C

with an applied temperature ramp of 1°C/3sec for the 35°C - 72°C transition.

7min	72°C

Remove half (20µl) of the amplification reaction and add to tube containing 4µl of 10x loading solution (50% sucrose, 0.2% Orange G in TE buffer).

Prepare 2% agarose gel (in TAE buffer), 0.5 µg/ml EtBr.

Load the samples on the gel plus molecular size marker.

Run the electrophoresis in TAE buffer, 0.5µg/ml EtBr, at 80 volts for 4 hours.

Place the gel on a UV transilluminaor,

wear protective glasses, turn on the UV light and make a photograph of the gel.

REFERENCES

1. Klein-Lankhorst, R.M., Vermun.A. Weide. R. Liharska, T. and Zabel, P. 1991 Theor. Appl. Genet. 83, 108
2. Rafalski, A.J., and Tigney, S.V. 1993, TIG 9,275
3. Williams, J.G.K., Kubelik. A.R.. Livak.K.J.. Rafalski, J.A., and Tingey S.V. 1990, Nucleic Acid Res. 18,6531.
4. Welsh, W., and Me Clelland, M. 1990, Nucleic Acid Res. 18,7213

35

SCARS AND CAPS, PCR-BASED MARKERS FOR GENETIC ANALYSIS

The Polymerase Chain Reaction (PCR) is an *in vitro* method of nucleic acid synthesis that enables the specific replication of a targeted segment of DNA (1-3). It is a simple automated procedure that facilitated many tasks in molecular biology research, for example, cloning and sequencing.

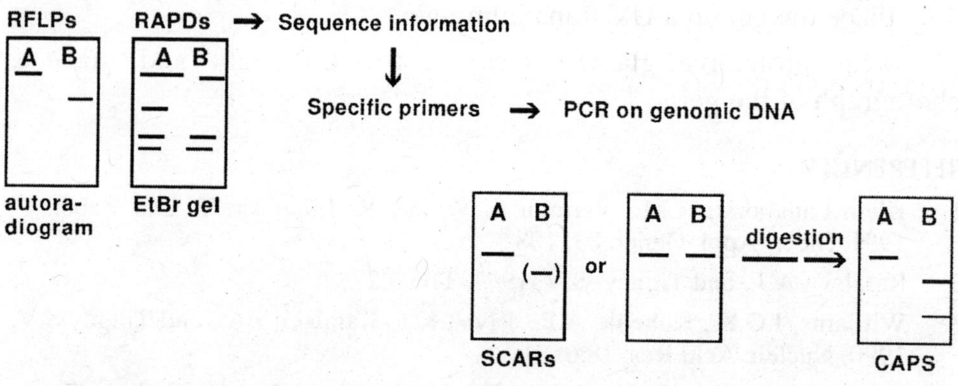

Fig. 35. Converting RFLP and RAPD markers into SCARs and CAPS.

Recently, techniques based on the PCR have been used to replace RFLPs and RAPDs markers in plant genetic analysis (5, 6). Sequence information RFLP and RAPD sequences with already known genome location is used to design specific primers (Fig. 35). Under stringent hybridization conditions these oligonucleotides prime the amplification of genetically defined loci, called Sequence Characterised Amplified Regions, SCARs, to on EtBr. stained agarose gel SCARs are visualised as single allele-specific bands, as presence/absence of the PCR products or as length variation (Fig.35.1). Very often at these conditions alleles direct amplification of nonpolymorphic products and additional polymorphisms must be searched for. Cleavage with restriction enzymes of the PCR products can reveal sequence polymorphisms between Cleaved Amplified Polymorphic Sequences CAPS (6). Both SCARs and CAPS are ideal marker for large scale screening of populations or genomic libraries (cosmids, BACs, YACs).

Here, SCARs and CAPS will be amplified that have been developed for chromosome 6 of tomato (*Lycopersicon esculentum*). These markers were derived from RAPD markers OPA 13 and OPC 8 (see the RAPD markers chapter). As templates, use genomic DNAs from *L. esculentum, L. penpellii* and the *L. esculentum* chromosome 6 substitution line WSL6.

SCAH13b gives a 475 bp fragment from *L. pennellii* and WSL 6, but no product from the *L. esculentum* line Moneymaker. CAPSC8b amplifies 400 bp single band on the three templates. After incubation with Hae III enzyme only the product from Moneymaker line is digested to 150 bp and 250 bp fragments.

Procedure

Precautions: Because PCR provides a method to produce many copies of a specific nucleotide sequence from minute quantities of DNA, precautions should be taken to avoid contaminations.

— autoclave solutions
— aliquot reagents
— wear disposable gloves
— avoid splashes
— use positive displacement pipettes
— premix reagents
— add DNA last
— choose positive and negative controls carefully

Stock solutions

Primers (Forward and Reverse) for SCAH13b and CAPS C8b 0.1 μg/μl

PCR buffer 10× (0.5M KCl, 0.1M Tris. Cl pH 8.3, 15mM $MgCl_2$, 0.1% gelatin)

dNTPs 1 mM for each dNTP

Taq polymerase (Perkin Elmer) 5U/μl

templates, genomic DNA from tomato, *Lycopersicon esculentum*

1. Moneymaker 0.1 μg/μl
2. *L. pennellii* 0.1 μg/μl

3. WSL6	0.1 µg/µl
*Hae*III enzyme (BRL)	1U/µl
buffer 2 (BRL)	

Amplification reacton

Prepare a mixture of the PCR reagents for each SCAH13b and CAPSC8b, and divide in four 0.5ml sterile microfuge tubes.

	in each tube	2 × 4 tubes
10× PCR buffer	5 µl	20 µl
1mM dNTPs	5 µl	20 µl
Primer F (0.1 µg/µl)	1 µl	4µl
Primer R (0.1 µg/µl)	1 µl	4 µl
Taq polymerase (5U/µl	0.2 µl	8 µl
sterile H_2O	36.8 µl	147.2 µl
Add for each SCAH13b and CAPSC8b		
in tube 1 Moneymaker DNA (100 ng)	1µl	
in tube 2 *L. pennellii* DNA (100 ng)	1µl	
in tube 3 WSL6 DNA (100 ng)	1µl	
in tube 4 sterile water	1µl	

Close the tubes, vortex and spin down.

Overlay the reaction mixture with 50µl of mineral oil (this prevents evaporation of the sample during repeated cycles of heating and cooling).

Transfer the samples in the PCR machine and carry amplification at following conditions:

5 min	94°C
	20 cycles
30 sec	94°C
30 sec	60°C
90 sec	72°C
7 min	72°C

1. Withdraw a half (25µl) of each amplification reaction and add to tube containing 4µl of 10× loading solution (50% sucrose, 0.2% orange G in TE buffer)
2. Prepare 1% agarose gel (in TAE buffer), 0.5µg/ml EtBr.

3. Load the samples on the gel plus molecular size marker
4. Run the electrophoresis in TAE buffer, 0.5 μg/ml EtBr, at 80 volts for 2 hours.
5. Place the gel on a UV transilluminator, wear protective glasses, turn on the UV light and make a photograph of the gel.

FOR CAPSC 8b only

1. Transfer the rest of the reactions (except the control) into clean Eppendorf tubes (1.5ml).
2. Add to each tube: 3 μl buffer 2, 12 μl *Hae*III enzyme.
3. Close the tubes, vortex and spin down.
4. Incubate 1 hour at 37°C.
5. Load the samples on the gel plus molecular size marker.
6. Run the electrophoresisin TAE buffer, 0.5 μg/ml EtBr. a 80 volts for 2 hours.
7. Place the gel on a UV transilluminator, wear protective glasses, turn on the UV light and make the photograph of the gel.

REFERENCES

1. Saiki, R.K. S. Scharf, F. Faloona, K.B. Mullis, G.T. Horn, H.A. Erlich, and N. Arnheim 1985. Science 230, 1350.
2. Mullis, K.,F. Faloona, S.Scharf, R. Saiki, G. Horn, and H- Erlich 1986. Cold Spring Harbor Symp. Quant. Biol 51,263
3. Mullis, K.B. and F. Faloona 1987. Methods Enzymol. 155, 335
4. Saiki R.K., D.H. Gelfand, S. Stoffel, S.J. Scharf, R. Higuchi, G.T. Horn, K.B. Mullis, and H.A. Erlich 1988. Science 239, 487.
5. Paran I, and R.W. Michelmore 1993. Theor. Appl. Genet. 85: 985-993
6. Konieczny, A. and F.M. Ausubel 1993. The Plant Journal 4:403-410

36

SITE-DIRECTED MUTAGENESIS

Oligonucleotide directed or *in vitro* site-directed mutagenesis is an invaluable technique used to induce specific alterations in cloned DNA, including base pair substitution, insertions and deletions. It is an important tool for studying protein structure-function relationships, gene expression, and vector modification. Stratagene's Chameleon™ doublestranded, site-directed mutagenesis kit allows site-specific mutation in virtually any doublestranded plasmid, thus eliminating the need for subcloning into M13-based bacteriophage vectors and for ssDNA rescue. (1)

The double-stranded target plasmid DNA is first heat denature, and two oligonucleotide primers are simultaneously annealed to one of the strands (2, 3, 4, 5, 6). One primer, the selection primer, changes one nonessential unique restriction site to a new restriction site. The other primer encodes for a specific mutation in a defined sequence of the gene of interest inserted in the plasmid. Both of these primers are extended around the plasmid using a nucleotide mix and an enzyme mix containing native T7 DNA polymerase and T4 DNA ligase. The newly created strand now contains the desired mutation but no longer contains the unique restriction site. The plasmid DNA is subsequently digested with the restriction enzyme that corresponds to the non-essential unique restriction site present in the unmutated parental plasmid. The unmutated parental plasmids are digested by the restriction enzyme, while the mutant plasmids remain undigested. The extended and digested DNA is then transformed into a repair-deficient mutS *Escherchia coli* strain which cannot distinguish the original unmuated strand from the newly created strand containing the desired mutation. The mutS-deficient strain randomly selects one of these strands as "correct" and changes the other strand to be complementary to the chosen strand. Since selection of the correct srand is random half of the plasmids contain the desired mutation and the other half convert back to the original plasmid. Because the transformation efficiency of circular plasmids is exponentially greater than linear plasmids, mutants are favoured.

Reagents and Media

SOB Medium (per litre)

20.0 g of tryptone

5.0 g of yeast extract

0.5 g of NaCl

Autoclave

Add 10 ml of 1 M $MgCl_2$ and 10 ml of 1M $MgSO_4$/liter of SOB medium prior to use

Filter sterilize

2× YT Broth (per Liter)

10 g of NaCl

10 g of yeast extract

16 g of tryptone

Add deionized H_2O to a final volume of 1 liter

Adjust to pH 7.5 with NaOH

Autoclave

LB-Ampicillin Agar (per Liter)

1 liter of LB agar

autoclave

Cool to 55°C

Add 50 mg of filter- sterilized ampicillin.

Pour into petri dishes (~ 25 ml/100-mm plate)

SOC Medium (per 100 ml)

SOB medium

Add 1 ml of a 2M filter-sterilized glucose solution or 2 ml of

20% (w/v) glucose prior to use.

Filter sterilize.

LB Agar (per Liter)

10g of NaCl 10g of tryptone

5g of yeast extract

20g of agar

Add deionized H_2O to a final volume of 1 liter

Adjust pH to 7.0 with 5N NaOH

Autoclave

Pour into petri dishes (~25 ml/100-mm plate)

LB-Ampicillin-Methicillin Agar (per Liter)

(Use for reduced satellite colony formation)

1 liter of LB agar

Autoclave Cool to 55°C

Add 20 mg of filter-sterilized ampicillin

Add 80 mg of filter-sterilized methicillin

Pour into petri dishes (~25ml/100 mm plate).

Protocol

Step I : Annealing the Primers to the DNA

Simultaneously anneal the two oligonucleotide primer-a selection and a mutagenic primer to the double-stranded target plasmid DNA by preparing the control and experimental reactions indicated below in 1.5-ml screw-cap microcentrifuge tubes:

1 µl (500 ng or 0.25 pmol) of pWhitescript plasmid DNA

5 µl (300 ng or 25 pmol) of *Kpnl* selection primer

5 µl (280 ng or 25 pmol) of blue mutagenic primer

2 µl of 10× mutagenesis buffer

7 µl of double-distilled water (ddH2O) to a final volume of 20 µl

1. Place the microcentrifuge tubes in a boiling water bath for 5 minutes and then immediately place the tubes on ice for 5 minutes.
2. Incubate the mocrocentrifuge tubes at room temperature (23-25°C) for 30 minutes.

Step II: Extending the Primers and Ligating the New Strands

For both the control and experimental reactions, extend the primers with T7 DNA polymerase and ligate the new strands with T4 DNA ligase as indicated below:

1. Prepare a fresh 1:10 dilution of the 10× enzyme mix in 9μl of enzyme dilution buffer.

   Note: It is crucial to use the fresh 1-10 enzyme dilution within 6-8 hours of preparation.

2. To each microcentrifuge tube, add 7μl of the nucleotide mix. The nucleotide mix must be added to the reaction first or the 3'-to 5'-nuclease activity of native T7 DNA polymerase will degrade the primers, Stir the 1:10 enzyme dilution with a pipet tip and add 3 μl of this fresh 1:10 enzyme dilution to the reaction. Thoroughly mix the reaction by pipetting the solution up and down and then spin briefly in a microcentrifuge.
3. Incubate the mocrocentrifuge tubes at 37°C for 1 hour.
4. Inactivate the T4 DNA ligase by incubating the mocrocentrifuge tubes at 70-80°C for 10-15 minutes to prevent religation of the digested strands during the subsequent digestion in Step III.

Step III: Digesting with the Restriction Enzyme for Selection

After the reactions cool, digest both the control and experimental reactions with a restrictin enzyme to eliminate those plasmids that did not anneal with the selection primer.

The reactions are now 30μl in 1 x mutagenesis buffer.

1. Add the following components to the control reaction:

   20U of *Kpnl* restriction enzyme

   ddH_2O to a total reaction volume of 60μl

   (Because *Kpnl* is active in 0.5× mutagenesis buffer, no buffer is added and the control reaction is diluted to twice its volume, diluting the buffer concentration from lx to 0.5×. For the requirements of other restriction enzymes, please consult a restiction enzyme buffer chart available from the respective manufacturer.)

2. Incubate the control reaction at 37°C for 1 hour.

Step IV: Transforming into XL*mutS* Competent Cells

1. Thaw the XL*mutS* competent cells on ice.
2. Gently mix the cells. Aliquot 90μl of the XL*mutS* competent cells into two prechilled 15-ml Falcon 2059 polypropylene tubes.

3. Add 1.5 μl of the β-mercaptoethanol to the 90μl aliquots of XL*mutS* competent cells to yield final concentrations of 25mM.
4. Swirl the contents of the Falcon 2059 polypropylene tubes gently. Incubate the cells on ice for 10 minutes, swirling gently every 2 minutes.
5. Add 1/10 of the volume (not to exceed 9μl) of the control and experimental reactions that have been digested with a restriction enzyme (see Step III) to each Falcon 2059 polypropylene tube and swirl gently. Use 6 μl for the control reaction.
6. Incubate the Falcon 2059 polypropylene tubes on ice for 30 minutes.
7. Heat pulse the Falcon 2059 polypropylene tubes in a 42°C water bath for 45 seconds. The length of time of the heat pulse is critical for obtaining the highest efficiencies.

   Note: Preheat the SOC medium (see Preparation of Media and Reagents) to 42°C at this step.
8. Incubate the transformation mixture on ice for 2 minutes.
9. Add 0.9 ml of preheated SOC medium and incubate the Falcon 2059 polypropylene tubes at 37°C for 1 hour with shaking at 225-250 rpm.
10. Plate 100 μl of the transformation mix on LB-ampicillin agar plates, containing X-gal and IPTG, to verify the Lac phenotype. For color selection, spread 20 μl of 0.2M IPTG and 20 μl of 10% (w/v) X-gal on LB-ampicillin agar plates 30 minutes before plating the transformants.

    Note: Do not mix IPTG and X-gal, since these chemicals will precipitate. X-gal should be prepared in dimethylformamide (DMF). Prepare IPTG in sterile H_2O.
12. Incubate the agar plates overnight at 37°C. Approximately 45-50% of the colonies should display the blue mutation.

Step V: Enriching for Mutant Plasmids

13. Enrich for mutant plasmids by adding the remaining transformation mixture that was not plated to 3 ml of 2× YT broth, supplemented with an appropriate antibiotic for the experimental plasmid.
14. Grow the culture overnight at 37°C with shaking.

Day 2

1. Perform a miniprep plasmid DNA isolation from 1.5 ml of the overnight culture.
2. Perform digestion of the miniprep DNA as follows:
   a. Digest 10 µl (=500 ng) of the resulting DNA isolation with the same selection restriction enzyme used in Step III. The digestion volume should be 10× the volume of added enzyme. Use at least 20 U of the restriction enzyme and digest under appropriate conditions for the selection enzyme. Control Reaction.

   10µl of miniprep DNA

   1 µl of 10× mutagenesis buffer (0.5× final concentration)

   20 U of *Kpnl* restriction enzyme

   ddH_2O to a final volume of 20 µl.

   b. Incubate the digestion for 1-2 hours or overnight (inorder to obtain a higher percentage of mutants) at 37°C or at the recommended temperature for the enzyme.

Step VI : Final Transformation into competent Cells

Transform the digested DNA into the *Epicurian coli* XLl Blue competent cells or into a cell line of choice as outlined below:

Transform 40 µl of *Epicurian coli* XLl-Blue competent cells with 1/10 the volume (2 µl) of the digested DNA.

Note: When aliquoting, keep the competent cells on ice at all times.

It is essential that the Falcon 2059 polypropylene tubes are placed on ice before the cells are thawed and that the cells are aliquoted directly into the prechilled tubes. Aliquot the unused cells in increments of 40pl and freeze for only one freeze-thaw cycle.

1. Add 0.68 µl of the β-mercaptoethanol provided with the kit to the 40µl of competent cells to yield a final concentration of 25 mM.
2. Swirl the contents of the Falcon 2059 polypropylene tube gently. Incubate the cells on ice for 10 minutes, swirling gently every 2 minutes.

3. Heat pulse the falcon 2059 polypropylene tubes in a 42°C water bath for 45 seconds. The length of time of the heat pulse is critical for the highest efficiencies.

   Note: Preheat the SOC medium to 42°C at this step.

4. Incubate the transformation mixture on ice for 2 minutes.
5. Add 0.45 ml of preheated SOC medium and incubate the Falcon 2059 polypropylene tubes at 37°C for 1 hour with shaking at 225-250 rpm.
6. Transformation mix is plated on LB-ampicillin agar plates, containing X-gal and IPTG, to verify the Lac phenotype. For color selection, spread 20μl of 0.2M IPTG and 20 μl of 10% (w/v) X-gal on LB-ampicillin agar plates 30 minutes before plating the transformants. If plating 5 μl, expect to see 150-300 colonies.
7. Incubate the plates overnight at 37°C. Approximately >70% of the colonies from the control reaction should display the blue mutation.

REFERENCES

1. Anon 1998. Training Manual "Recent Techniques in Plant Molecular Biology and Biotechnology". NRC on Plant Biotechnology IARI, New Delhi.
2. Carter, P. 1987. Method Enzymol. 154: 382-403
3. Kramer, W. and Firtz, H.J. 1987. Method Enzymol. 154: 350-367.
4. Kunkel, T.A. 1985. Proc. Natl. Acad. Sci. USA., 82: 488-492 Kunkel.
5. Robert,T.A., J.D. and Zakour, R.A. 1987. Method Enzymol, 154: 367-382.
6. Lewis, M.K. and Thompson, D.V. 1990. Nucleic Acid Res., 18: 3439-3443.
7. Ray, F.A. and Nickoloff, J.A. 1992. Biotechniques, 13:342-346.
8. Zoller, M.J. and Smith, M. 1987 Method Enzymol, 154: 329-350.

37

ISOLATION OF RNA AND NORTHERN BLOTTING

ISOLATION OF RNA

Solutions and Buffer

(1) Denaturing solution (Sol. D)

(4M guanidinium thiocyanate, 25 nM sodium citrate pH 7.0, 0.5% sarcosyl, 0.1 m 2-mercaptoethanol).

guanidinium thiocyanate	50g
Sterile water	58.6ml
(pH 7.0)	
10% sacrosyl	5.28ml
Make up the volume to	100 ml

(This solution can be stored for at least 3 months at RT. Do not autoclave or filter sterilize the solution. Prepare all the stocks and sterilize. Use sterilized beaker, spatula and stirbar for preparing the solution.

0.36 ml of 2-mercaptoethanol/50 ml of stock solution is added just before use.

(2) For 10 ml of 2M sodium acetate (pH 4.0)

Sodium acetate-1.64g

Dissolve in minimum amount of distilled or milli Q water. Adjust pH to 4.0 with Acetic acid and make up the volume to 10 ml. Autoclave and use.

(3) For 10 ml of 0.75 M sodium citrate (pH7.0)

Tri-sodium citrate 0.22g

Dissolve in minimum amount of distilled or milli Q water. Adjust pH to 7.0 using HCl and make up the volume to 10 ml. Autoclave and use.

(4) 10% sacrosyl

N-lauryl
Sacocine 10gm
St. H_2O to 100 ml
Do not autoclave or filter sterilize.

Protocol

1. Weigh 3 g of cowpea seeds or any other tissue. Homogenized with 15ml of solution D using polytron.
2. Transfer the homogenate into a 50 ml sterile polypropylene tube and mix by inversion.
3. Add 1.5 ml of 2 M Sodium acetate (4.0). Mix thoroughly by inversion.
4. Then add 15 ml of Phenol (water saturated) and 3ml of chloroform-isoamyl-alcohol mixture (49:1). Again mix thoroughly by inversion.
5. Shake the final suspension vigorously for 10 seconds and cool in ice for 15 min.
6. Spin the samples at 10,000 × g for 20 min at 4°C. (After centrifugation, RNA is present in the aqueous phase, whereas, DNA and proteins are present in the interphase and phenol phase).
7. Transfer the aqueous phase to another fresh sterile tube and mix with 15ml of isopropanol and place at 20°C for at least 1 hr. to precipitate RNA.
8. Sediment RNA at 10,000 g for 20 minutes.
9. The resulting RNA pellet is dissolved in 3 ml of Sol. D and re-precipitate with 1 vol. of isopropanol at -20°C for 1 hr.
10. Centrifuge again for 10 min at 4°C at 10,000 × g.
11. Wash the pellet with 75% ethanol by resuspending in it.
12. Sediment again, vacuum dry the pellet and dissolve in 750 μl of 0.5% SDS at 65°C for 10 min.
13. Determine concentration at 260 nm spectrophotometrically.
    Relationship IOD = 40 μg RNA

ISOLATION OF TOTAL RNA BY GTC METHOD

Reagents

Solution A : 4 M GTC

50 g GTC (Guanidine thiocynate)

2.5 ml 1 M Sodium citrate pH 7.0

0.7 ml b-ME

Adjust pH to 7.0

Make up Vol. to 100 ml (Do not authoclave or filter). Use autocloaved milli-Q water for every solution.

Solution B : CsCl cushion solution

48g CsCl

1 ml of 0.5M EDTA (pH 7.0)

Vol. to 50 ml (autoclave)

Solution C : RNA suspension buffer

Autoclave and keep for further use.

Make stocks and use

	Stock	Final conc.
1 M Tris-Cl (pH 7.5)	0.5 ml	0.01M
0.5M EDTA (pH-8.0)	1.0 ml	0.01M
20% SDS	25 ml	0.01%
5M NaCl	3.5ml	0.35M
St. H_2O	44.975ml	50.00ml

Dispense in aliquots and store at-20°C.

Protocol

1. Sterilize every glassware, stir bars, centrifuge tubes, Eppendorf tubes, tips, spatulas, homogenizer cup. Use gloves (disposable) for every step.
2. Take 10 ml GTC/lg of material(seed, embryos) in a homogenizer cup and homogenise in cold for 20 sec. each × 4 times.
3. Pass through 4 layers of cheese cloth.
4. Centrifuge for 15' at 10 K at room temperature.
5. Measure the volume and add CsCl at the rate of 3g/8ml.
6. To a 40ml centrifuge tube ($SW_{27.1}$ tube), add 6 ml of Soln. B and layer the supernatant carefully. Fill to the top, balance and layer with Soln. A if necessary.
7. Spin at 27K or 28K for 23 hrs. at 25°C.
8. Remove the GTC layer carefully without disturbing the pellet.
9. Invert the tube on tissue paper and remove any traces of GTC.
10. Cut the tube from 1 cm of bottom (necessary step).

11. Slowly add RNA suspension buffer (100 μl - 150 μl each time). Break the pellet and transfer to eppendorf tube.
12. Preheat water bath at 60°C.
13. Resuspend RNA in about 700-800 μl of Soln. C.
14. Heat the tubes and then keep in ice to cool and centrifuge at 4°C for 10-15sec. in Eppendorf centrifuge.
16. Take the supernatant in another tube and label it as #1.
17. Re-extract the pellet with Soln. C and heat at 60°C, vortexing often in between for 2 hr. Centrifuge and keep supernatant in another tube. Do not mix the two supernatants. Label it as # 2 separately.
18. Extract the supernatants with equal volumes of phenol. Chloroform (1:1 v/v equilibrated in 0.1M Tris pH 8.0. Store at 4°C for 1-2 times.
19. Re-extract the organic phase, if necessary (Optional).
20. Take the aqueous phase and add 2 vol. of 100% cold ethanol and keep at -20°C overnight (Do not use -70°C for precipitation).
21. Spin in Eppendorf centrifuge for 15 min. for pelleting RNA.
22. Wash the pellet with 70% chilled ethanol and centrifuge.
23. Remove the supernatant. Dry down the pellet and dissolve in sterile water.
24. For O.D. take 1 μl and dilute to 400 μl in 1 ml cuvette with sterile water. (Sterile the cuvette by immersing in a soln. of 0.1% SDS and proteinase K in sterile water and keep for overnight. Wash with sterile water. This could be omitted if RNA solution is not to be recovered).
25. Soak the gel casting tray, comb, buffer containing cells in Proteinase K (Pinch), ~5 mg/100 ml and 0.1% SDS in sterile water overnight (preferable) otherwise minimum for 4-5 hrs. rinse them in sterile water properly and then use for casting of agarose gel.

TOTAL RNA ISOLATION USING TRIZOL REAGENT

A. REAGENTS

1. TRIZOL reagent (GIBCO BRL, Life Technologies, Incl., USA)
2. Chloroform
3. Isopropanol
4. 75% Ethanol
5. RNase-free water

6. DEPC (diethyl pyrocarbonate)

B. ISOLATION

I. Homogenization:

Homogenise 100 mg leaf tissue in 1 ml of TRIZOL reagent.

II. Phase Separation

Incubate the homogenized sample for 5 min. at room temperature to permit the complete dissociation of nucleoprotein complexes. Add 0.2 ml of chloroform, cap the sample tube and shake vigorously by hand for 15-20 seconds, and incubate at room temperature for 2-3 minutes. Centrifuge the samples at 12000 × g for 15 minutes at 4°C. The mixture separates into lower red, phenol-chloroform phase, an interphase, and a colourless upper aqueous phase. RNA remains in the aqueous phase.

III. RNA precipitation

Transfer the aqueous phase to fresh tube and add 0.5 ml isopropanol to precipitate the RNA. Incubate samples at room temperature for 10 min. and centrifuge at 12,000 × g for 10 min. at 4°C.

IV. RNA Wash

Wash the RNA pellet once with 75% ethanol (500μl) by centrifugation at 7,5000 × g for 10 minutes at 4°C.

V. Redissolving the RNA

Dry the RNA pellet briefly in air. do not dry the RNA by centrifugation under vacuum. Dissolve the RNA in RNase-free water, and incubate for 10 minutes at 55° to 60°C.

ISOLATION OF POLY (A)+ RNA

Solutions

A. Low salt buffer (LSB) (10 mM Tris, pH 7.5)

1 M Tris-Cl pH 7.5	1.0 ml
St. H_2O	99 ml

B. Medium salt buffer (MSB) (10mM Tris, pH 7.5, 0.1 M NaCl)

1M Tris-Cl pH 7.5	1.0 ml
5 M NaCl	2.0 ml
St. H_2O	97.0 ml

C. High salt buffer (HSB) (10 mM Tris, pH 7.5, 0.5 M NaCl)

1M Tris-Cl pH 7.5	1.0 ml
5M NaCl	10.0 ml
St. H_2O	89.0 ml

D. Loading buffer (2xHSB) (10mM Tris, pH 7.5, 1.0 M, Nacl)

1 M Tris-Cl pH 7.5	2.0 ml
5 M NaCl	20.0 ml
St. H_2O	78.0 ml

1. Autoclave buffer, column and tubing to be used.
2. Add SDS to 0.1% (1 ml of 20% SDS/100ml) in loading buffer SDS is not autoclavable.

Chromatography

3. Pack the column with oligo (dT) to 1 ml Vol.
4. Wash with 20 ml HSB
5. Apply total RNA (Take about 6 mg of total RNA, dilute to 1 ml with sterile water. Heat at 60°C in water bath for 5 min. Cool in ice and add equal vol. of loading buffer).
6. Collect the flow through and load again on column.
7. Repeat Step 6 once more
8. Wash with HSB ~ 20 ml
9. Wash with MSB ~ 20 ml
10. Keep the flow rate low and add LSB and collect 11-12 ml of sample flow through in SW 27.1 sterile tube.
11. Add 5 M NaCl to make it 0.2M.
12. Add 2 vol. of 100% ethanol and keep at -20°C for overnight. Do not keep in -70°C straight.
13. Centrifuge RNA at 25 K for 2.30 hrs at 2°C.
14. Remove the supernatant and let dry the tube.
15. Dissolve the pellet in appropriate amount of sterile water.
16. Check O.D. at 260 nm and 280 nm.

ELECTROPHORESIS OF RNA

Solutions

		Final conc.
(A) 10× MOPS	**pH 7.0 (1 litre)**	
41.86 g	MOPS (M.W. 209.3)	0.2 M

6.8 g	NaOAC (M.W. 136.1) (Sodium Acetate)	50mM
3.362 g	EDTA	10mM

Adjust pH to 7.0 (autoclave)
Soln. turns yellow after autoclaving.

(B) Loading Buffer (5×)

Final conc.

Glycerol	10 mL	50%
0.45 M EDTA	40 μl	1 mM
0.2% BPB	0.04 g	
0.2% xylene cresol	0.04 g	
	Total	**20ml**

Protocol

1. Preparation of gel (agarose gel-1%)

   To 1.2 g agarose add 104.4 ml sterile water and boil. Then add 12ml 10xMOPS (Soln. A) 3.6 ml formaldehyde. Allow it to cool to 55°C and then cast the gel.

2. Preparation of sample

   4 μl RNA (if concentrated use less and make up with water). 1.56 μl 10× MOPS [3-(N-morpholino) propane sulfonic acid)], 3.12 μl formaldehyde.

   8.88 μl deionised formamide.

   Heat at 60°C for 7.5 min.

   Cool in ice for 5 min.

   Add 4 μl 5× loading buffer.

3. Load the sample on gel and run at 20V for overnight (40V for big gel).
4. Stain the gel with EtBr. (1μg/ml of sterile water). Shake for 10 min.
5. Destain with several changes in sterile water. Take picture. Two bands 18S and 28S of rRNAs should be visible.

NORTHERN BLOTTING

Solutions and Buffers

1 M NaOH; 20 X SSC; Hybond-N^+ membrane.

Protocol

1. Remove formaldehyde from the gel by repeated washing with sterile water and then soak in excess volume of 0.05 M sodium hydroxide for a brief period.
2. Wash the gel and soak in 20× SSC for 45 min.
3. Remove the gel from 20× SSC and transfer the RNA molecules to nylon membrane (as described for Southern blotting).
4. After overnight transfer, remove the membrane, blot mark for orientation and wash in 6× SSC.
5. Air dry the membrane, keep in a sealed bag.

Prehybridization solution

50% deionized formamide

5× Denhardt's solution

5× SSC

0.1%w/vSDS

Denatured salmon sperm DNA at a final concentration of 100 μg/ml of the solution.

NORTHERN HYBRIDIZATION

Buffers and Solutions

Pre-pybridization buffer: Radiolabelled DNA probe.

Protocol

1. Transfer the filter to a bottle and add prehybridization solution (100 μl/cm2). Incubate at 65°C for 30 min.
2. Denature the radiolabelled probe and add to the pehybridization buffer.
3. Allow hybridization by gentle shaking in hybridization oven in hybridizations bottle or box at 65°C for 18h.
4. Conduct washing and autoradiography next day.

DEVELOPMENT OF AUTHORADIOGRAPHS

Buffers and Solutions

X-ray film developer : X-ray film fixer.

Protocol

1. In the dark room, place sequentially the tanks for X-ray film developer (No. 1), water (No. 2) and X-ray film fixer (No. 3).

2. Carefully remove the X-ray film (s) exposed to the hybridized gel/filter (s) and dip in the developer solution. (Now you can switch on the safe light).
3. Develop for suitable time period (check occassionally, at regular intervals) wash in water and fix in the fixer solution for at least half an hour.
4. Wash off the completely in running tap water and air dry the film.

REFERENCES

1. Brawerman, G, Medecki, J. and Lee, S.Y. 1972. Biochemistry 11: 637.
2. Cashmore, A.R. 1982. In: Methods in Chloroplast Molecular Biology (Edelman, M., Hallick, R.B. and Chua, N.H. eds.) Elsevier, Amsterdam, pp 387-392.
3. Davies, E., Larkins, B.A., and Knight, R.H. 1972. Plant Physiol. 50: 581-584.

38

ISOLATION OF CHLOROPLAST DNA

In addition to the genetic information contained in the nucleus, plant cells also contain DNA in their chloroplast and mitochondria (Schuler and Zielinski, 1989). The DNAs of these organells resemble both bacterial and eukaryotic nuclear DNAs in their organization. Unlike mitochondrial DNA (about 1%) the chloroplast DNA (Plastid DNA) represent about 30% portion of the DNA in a mature leaf cell (Schleif and Wensink, 1981, and Palmer, 1985). Recently, the complete nucleotide sequences of chloroplast DNA from a liverwort (Ohyame *et. al.*, 1986) and an angiosperm (Shinozaki *et. al.* 1986) have been determined. Fig 38 shows a genetic map of tobacco chloroplast DNA which contain over 120 genes, plastid genome.

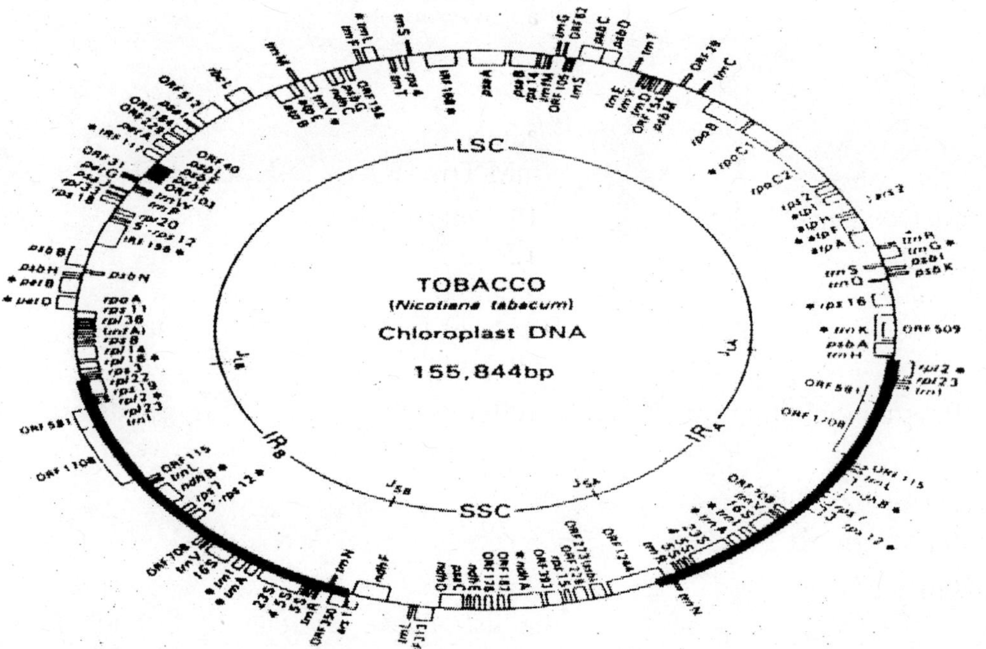

Fig. 38. Chloroplast/Plastid genome of tobacco showing different genes.

A detailed procedure for chloroplast DNA isolation, its restriction digestation, southern blotting and hybridization, and production of autoradiogram is described in this chapter (Schuler and Zielinsky, 1989 Herrmann 1982; Kolonder and Tiwari, 1979, Krebbers *et al.*, 1982; McIntosh *et. al.* 1980, Palmer, 1982; 1983; and Palmer and Thompson, 1981).

Wash buffer

350 mM sorbitol	6.4 g sorbitol
50mM Tris-HCl (pH 8.0)	5.0 ml 1M Tris-HCl (pH 8.0)
20 mM EDTA	10ml 200 mM EDTA adjust final volume to 100 ml with distilled water autoclave
10mg/ml proteinase K solution	100 mg proteinase K 10 ml wash buffer Store at -20°C

Sterilre 10-ml graduated cylinders (or test tubes) Sterile plastic test tubes for high-speed centrifuge Sterile, siliconized 15-ml corex test tubes

0.672g/ml CsCl stock solution	26.88 CsCl adjust final volume to 40 ml with 50 mM Tris-HCl (pH8), 20 mM EDTA
1.24g/ml CsCl stock solution	248 g CsCl adjust final volume to 200 ml with 50 mM Tris-HCl (pH8), 20 mM EDTA
CsCl "balance" solution	19.0 ml wash buffer 1.5 ml 20% sarkosyl 5.0 ml 0.672g/ml CsCl 33.5 ml 1.24g/ml CsCl 49.0 ml
10mg/ml ethidum bromide stock	100 mg ethidium bromide adjust final volume to 10 ml with sterile water

Tris-EDTA buffer

50mM Tris-HCl (pH 8.0)	6.06 g Tris base
20mM EDTA	7.44 g Na2EDTA $2H_2O$ adjust final volume to 1 liter with distilled water dissolve in 900 ml distilled water

	dissolve in 900 ml distilled water adjust pH to 8.0 with HCl autoclave
Buffer-saturated isopropanol	15ml TE buffer 10g CsCl dissolve CsCl in TE buffer, then add 100 ml isopropanol, isopropanol should be buffer saturated for at least 24 hours.
1mMEDTA	0.37 g Na_2 EDTA-$2H_2O$ dissolve in 900 ml distilled water adjust pH to 8.0 with HCl adjust final volume to 1 liter with distilled water autoclave
TE.1 buffer	
10mM Tris-HCl (pH 8.0)	1.21 g Tris base
0.1 mM EDTA	0.04 g Na_2 EDTA-$2H_2O$ dissolve in 900 ml distilled water adjust pH to 8.0 with HCl adjust final volume to 1 liter with distilled water autoclave 11.7g NaCl
2M NaCl	adjust final volume to 100 ml autoclave

Sucrose Gradient Method for the Isolation of chloroplasts

Protocols

1. Use 20 g of fresh leaf tissues, try to destarch material by storing plants in the dark 1-2 days before preparation. If leaf material is removed from plant prior to prep, store in plastic bag with wet paper towels in the dark and in the refrigerator.
2. Remove young growing leaves from plants for the preparation. Excise away nongreen and necrotic tissue. Rinse in cold water several times; rinse in 50 ml cold 10% Clorox 1-2 minutes; rinse extensively with cold water, being sure to rinse off bleach. Blot tissue dry with towels or use salad spinner. Cut into small pieces if necessary.
3. Repair to the cold room. Blend tissue in 5 volumes/gram fresh weight of extraction buffer (100 ml) in a waring blender. Use two or three 5 second bursts on low speed.

4. Filter rapidly through six layers of cheese cloth and one layer of Miracloth (gravity filtration).
5. Pour filtrate into 250 or 500 ml centrifuge bottles; spin at 250 rpm in GSA, GS3, or HB4 for 15 minutes at 4°C.
6. Discard supernatant and add 8 ml of wash buffer. Resuspend green pellet gently with fine camel's hair brush, swirling contents in centrifuge bottle. Do not vortex.
7. Layer resuspended organelles over a 30:45:60% sucrose step gradient. Top off gradient with wash buffer. Balance tubes. Centrifuge in the SW27 (Beckman) or AH637 (Sovrall) rotor at 25,000 rpm for 60 minutes at 4°C.
8. Use a wide-bore pipet or cut-off pipet tip to remove intact chloroplast band, the one at or closest to the 30:45% interphase. It should be very dark green.
9. Estimate volume of chloroplast band. Add 3 volumes cold wash buffer. Centrifuge at 4000 rpm for 10 minutes at 4°C in the SS_{34} rotor. Resuspend pellet in 9 ml wash buffer and hold on ice or in refrigerator until you need the chloroplasts.

If glassware is used when working with small amount of nucleic acid, it should be siliconized to minimize loss of nucleic acid due to binding to the glass. Alternatively, plasticware (poly propylene and poly styrene) can be used without silanization.

1. Bring chloroplast sample up to 2.7 ml with wash buffer.
2. Add 0.3 ml of 10 mg/ml proteinase K solution. Mix gently by inversion; incubate at room temperature for 15 minutes.
3. Add 0.3 ml of 20% sarkosyl.. Mix by gentle inversions. Once sarkosyl has been added, the solution should be handled very gently in subsequent steps of minimize shearing of the high molecular weight DNA. Incubate the mixture at room temperature for 30-45 minutes, gently inverting the solution about 30-50 times during this period.
4. Add 1.0 ml of a 0.672 g/ml CsCl stock solution. Mix with gentle inversions (about 30-50 times) and allow the solution to stand 1 hour (or overnight) at 4°C.
5. Pellet membranes and starch particles by centrifuging sample for 20 minutes at 12,000g (10,000 rpm in a Beckman JA17 rotor).
6. Carefully decant DNA into a 10-ml graduated cylinder or test

tube. Adjust volume to 4.8 ml per gradient with Tris-EDTA buffer. For each gradient, add 6.7 ml of the 1.24 g/ml CsCl stock solution and 0.3 ml of the 10 mg/ml ethidium bromide (EtBr) solution. Invert slowly to mix solutions. After EtBr has been added to the DNA, the DNA-EtBr solution should be handled in subdued light to minimize light catalyzed nicking of the DNA.

7. To assemble CsCl-EtBr gradients
   (a) Transfer the 11.8 ml of DNA-CsCl solution to a polycarbonate ultracentrifuge tube.
   (b) Attach the ultracentrifuge tubes and balance with CsCl "balance" solution.
   (c) Attach the cap assemblies to the tubes and fill tubes completely with CsCl balance solution; balance tubes again and screw the cap assemby closed a small air bubble in the centrifuge tube will not cause the tubes to collapse.

   Note: "Quik-Seal" test tubes can be used instead of polycarbonate ultracentrifuge tubes with cap assemblies. When using Quik-Seal tubes, a metal adapter should be placed over the tube during ultracentrifugation, otherwise the tube will collapse.

8. Centrifuge the tubes in the Beckman 50 Ti rotor at 40,000 rpm for 40 hours, or in the 70.1 Ti rotor at 53,000 rpm for 10 to 20 hours, at 20°C.

   Note: Low temperature (4°C) will cause the CsCl to precipitate out of solution and change the CsCl density in the gradient. Consequently the DNA will band too low in the centrifuge tube.

9. The relaxed and supercoiled DNA bands present in the gradient are visualized with long-wavelength UV light, which minimize nicking of the DNA.
   (a) Secure centrifuge tube to a ringstand with a clamp and remove the set screw. If using Quik-Seal tubes, puncture top of the test tube to prevent a vacuum from forming in step d below.
   (b) Visualize DNA bands with a 360 nm UV source. Usually the chloroplast DNA is present in the relaxed and linear forms.
   (c) If using polycarbonate ultracentrifuge tubes, expel air from a sterile Pasteur pipet, gently move it down through the gradient and position it just under the DNA band. Withdraw a

minimal volume of DNA and transfer it to a plastic 15ml or 50 ml screw-cap test tube.

d. If using Quik-Seal tubes, lightly smear acetone over the area of the tube where the DNA has banded. Wipe dry, then put a piece of sticky medical adhesive tape there. Lightly smear the adhesive tape with acetone and wipe dry. Carefully puncture the tape with a needle and syringe (18-or-16-gauge large-bore needle). Do not position your hand on the test tube wall opposite the point of needle insertion, withdraw a minimal volume of DNA using a 5 ml syringe. Very gently expel the DNA from the syringe into a plastic 15ml or 50 ml screw-cap test tube. If relaxed and supercoiled DNAs are being collected separately, carefully withdraw the needle and seal the hole with a second piece of adhesive tape.

10. *Note* the original volume of the CsCl-banded DNA solution (usually 1 to 2ml). Gently extract the DNA with an equal volume of buffer-saturated isopropanol (isopropanol is micible with water but not with a concentrated CsCl solution). Gently invert the test tube several times and then allow the phases to separate by gravity. The ethidium bromide should partition into the upper (isopropanol) phase, which can be removed with a Pasteur pipet and discarded in the appropriate waste container (EtBr waste should be chemically inactivated before disposal). The aqueous phase should be reextracted with isopropanol until the aqueous phase is no longer pink (usually three to six extractions altogether). After the final extraction, adjust the volume of the DNA-CsCl solution to its original level using sterile water (the cloudiness of the solution should clear up). Transfer the aqueous phase to a (siliconized) 15 ml Corex test tube, being careful not to transfer any of the interface material.

11. Estimate the DNA volume and precipitate the DNA by adding.

    1. volume of TE. 1 buffer (to dilute out the CsCl)
    2. volumes ethanol (final conc. = 50% EtOH)

12. Chill the solution at -20°C for at least 4 hours and preferably over-night. (Do not chill at -70°C because the CsCl will precipitate out.)

13. Pellet the DNA by centrifugation at 12,000g (10,000 rpm in a Beckman JA17 rotor) for 30 minutes. Slowly decant the supernatant and let the Corex tube drain upside down in a test tube rack. The pellet can be left to air dry for 15-60 minute in this

fashion. Save the pellet. Dissolve the DNA in 400 µl sterile TE. 1 buffer. Allow the DNA to resuspend for at least 4 hours at 4°C (preferable overnight) Store the DNA at 4°C in microfuge tubes. (High molecular weight DNAs take a long time to resuspend so it's best to store them overnight before going on. Vacuum drying accentuates the problems associated with resuspending high molecular weight DNA.)

14. A second ethanol precipitation of the chloroplast DNA is usually required to fully remove residual CsCl which can interfere with subsequent restriction enzyme digestions. The second ethanol precipitation can be performed in a 1.5 ml microfuge test tube with 0.1 volume of 3 M sodium acetate and 2.0 volumes of 100% ethanol.
15. Resuspend final DNA pellet in 100 µl of TE.1. You will have too little DNA to measure with the spectrophotometer. In stead, the amount of DNA can be estimated on a gel, relative to a known amount of *Hind* III-cut λ DNA.

A. Restriction Digestion of Chloroplast DNA

1. Set up 40µl restriction digestions

   4µl 10× Eco Rl buffer
   20µl chloroplast DNA
   15µl sterile water
   1µl 5 U/µl Eco R1
   40 µl

   4 µl 10× *Pst* 1 buffer
   20 µl chloroplast DNA
   15 µl sterile water
   1 µl 5 U/µl *Hind* III
   40 µl

   4 µl 10× *Bam* H1 buffer
   20 µl chloroplast DNA
   15 µl sterile water
   1 µl 5 U/pl *Bam* Hl
   40 µl

2. Incubate restriction digestion for 45 minutes at 37°C. At the end of this period, add 8 µl 10 M urea loading dye. Load 24 µl of sample

per lane on gel. Don't forget to load the ^{32}P-labeled *Hind* III-cut 1 DNA standard.

3. Pour a 0.8% agarose gel as follows:
   (a) Dissolve 1.2g agarose in 150 ml lx agarose gel buffer.
   (b) heat in microwave or on hot plate until completely dissolved; cool slightly; pour into gel plate; adjust 13-well comb.
   (c) load two sets of samples onto your gel (because you will probe the Southerns with two different probes); the order should be *Hind*-cut λ DNA, *Eco* R1 DNA, *Pst* 1 DNA, *Hind* III DNA, *Bam* HI DNA, two spaces, Hind-cut λ DNA, *Eco* R1 DNA, etc.

   Note: Use lower percentage agarose gels for mapping the chloroplast DNA because these gels separate the higher molecular weight fragments better.
4. Run gel either at 20V overnight or at 100 V for 4 hours until bromophenol blue dye reaches the bottom of the gel.
5. Stain the agarose gel with 0.5 μg/ml EtBr in water for 30 minutes. Photograph gel.

B. Restriction digestion of Chloroplast DNA and Genomic DNA Southern Transfers

10 × *Eco* Rl buffer

10 × *Pst* l buffer

10 × *Bam* HI buffer

Sterile water

10M urea loading dye

^{32}P-Labeled *Hind* III-cut λ DNA standards

2× hybridization buffer	
10× SSC	88g NaCl (1.5M)
	44g sodium citrate (0.15M)
0.4% sarkosyl	4g sarkosyl
2× Denhardt's solution	
0.04% BSA	0.4 g BSA
0.04% polyvinyl-pyrrolidone	0.4 g polyvinylpyrrolidone
0.04% Ficoll	0.4 g Ficoll
	adjust final volume to 1 liter with distilled H_2O

0.25 M HCl	40ml conc. HCl adjust final volume to 2 liters with distilled water
0.5 M NaOH	40 g NaOH
1 M NaCl	116 g NaCl adjust final volume to 2 liters with distilled water
0.5 M Tris (pH7.5)	121 g Tris base
3 M NaCl	350 g NaCl add 1.8 liters water add conc. HCl to pH 7.5 adjust final volume to 2 liters with distilled water
20× SSC	
3 M NaCl	350 g NaCl
0.3 M sodium citrate	176 g sodium citrate adjust final volume to 2 liters with distilled water

C. Genomic DNA Southern Transfer (Southern, 1975)

1. Soak gel in 250 ml (enough to cover gel) 0.25 M HCl for 15 minutes with occasional swirling; drain; repeat a second time with 0.25 M HCl for 15 minutes. Acid treatment will depunnate DNA.

   Notes: * Be very careful about the times for the HCl and later washes. Too long an HCl treatment will cleave the DNA into tiny fragments which cannot stick to nitrocellulose; long NaOH and Tris washes will allow the smaller DNAs to diffuse out of the gel.

   * The layers of paper and towels over the slab gel must not contact the wet paper below because this will short circuit the liquid passing through the gel, nitrocellulose, and into the paper towels.

   * The nitrocellulose filters can be marked with India ink or permanent marker pen before putting them on top of the agarose gel.

   * If the filters are baked too long or at too high a temperature they become too brittle to work with be careful.

2. Rinse gel once briefly with distilled water.

3. Soak gel in 250 ml 0.5 M NaOH, 1 M NaOH, 1 M NaCl for 15 minutes; drain. Base treatment will cleave DNA at apurinic residues and denature the DNA fragments prior to transfer.
4. Soak gel in 250 ml 0.5 M Tris-HCl (pH 7.5), 3 M NaCl for 5 minutes; drain. Repeat with 0.5 M Tris-HCl (pH 7.5), 3 M NaCl for 15 minutes; drain.
5. Set the gel up for transfer to nitrocellulose paper.
6. Allow DNA to transfer to nitrocellulose filter for 3 hours or overnight at 25°C. The agarose gel should be paper thin after 3 hours. Cut the filter in half since you will hybridize each half with a different probe. Place filters on a paper towel and dry for 1 hour at 65-70°C (no higher) in a vacuum oven.

Southerns set up this way should be allowed to transfer for 2 hours or overnight at room temperature and then bake for 1 hour at 65-70°C.

D. Nick Translation

This procedure is the same as in Chapter 31. You will be provided with 0.5 μg/μl of a 593-bp *Bgl* I. Pst I fragment from pZmc 702, a subclone of the maize ATPase B-subunit gene and a 567-bp *Pst* I fragment from pZmc 701.

a subclone of the maize large subunit RuBis Co gene.

1. Set up the nick-translation procedure as follows:
   10 μl DNA (0.2-2.0 μg purified DNA fragment)
   10 μl 10x nick-translation buffer [500 mM Tris (pH 7.80).
   50 mM $MgCl_2$, 1 mg/ml BSA, 10 μM 2-mercaptoethanol]
2. μl of 20 mM dCTP, dGTP, dTTP stock
   2 μl [^{32}P] dATP (10 μ ci./μl 3000 ci/m mol)
   74 μl H_2O
   1 μl 1/10000 dilution of 1 mg/ml stock DNase
   (dilute immediately before use)
   1 μl DNA polymerase 1 holoenzyme
   For 100 μl reaction
   Incubate for 1 hour at 15°C in an ice bucket water bath behind isotope shield.
2. Set up a 4-ml G-100 Sephadex column to separate incorporated and unincorporated ^{32}P. Stopper a plastic 5 ml pipet with siliconized glass wool, fill pipet with TE buffer and add G-100 Sephadex until column bed coulmn=4 ml. Keep buffer on top of

the Sephadex so that column does not run dry. After it has adsorbed to the column, start eluting with sterile TE buffer. Collect the first blue peak that comes off the column, this excluded peak contains the incorporated radioactive nucleotides. Dispose of the unincorporated nucleotide (the second blue band) and the column in the radioactive waste area.

3. Count 1 μl of the probe on a paper circle in a scintillation vial without scintillant on ^{3}H channel. You should get 5.000-20,000 cpm/μl when purified DNA fragment is used for nick translation.

E. Hybridization of Southerns

1. Seal nitrocellulose filter into seal-a -meal bag; check to make sure that the seals are tight.
2. Mix prehybridization solution as follows:

   7.5 ml 10× SSC, 0.4% sarkosyl, 2× Denhardt's solution 6.0 ml formamide (final conc. = 40% formamide)

   1.5 ml H_2O

   15.0 ml
3. Cut off one corner of bag. Add prehybridization solution to bag. Reseal.

   Incubate for 10 minutes on your benchtop. Squeeze out excess liquid. (Usually the prehybridization step is performed for 4 hours to overnight to minimize back ground hybridization.)
4. Add 500,0000 cpm of each probe to a separate test tube containing 1 ml of sterile water. Boil the ^{32}P nick-translated DNA for 3 minutes in a boiling water bath to separate the DNA strands.

   Mix the heated probe with the hybridization solution (outlined below) and add to the seal-a-meal bag containing the filters. Try to get rid of as many bubbles as possible before resealing the bag.

   Set up each bag of filters with 15 ml of hybridization solution as follows:

   7.5 ml 10× SSC, 04% sarkosyl, 2× Denhardt's solution

   6.0 ml formamide

   0.5 ml H_2O

   1.0 ml ^{32}P nick-translated DNA in H_2O (use approximately 500,000 cpm/filter) 15.0 ml

   Hybridize filters for 12-20 hours (overnight) at 40°C. The relatively low hybridization temperature is due to the following :

(1) the 37-38% GC content of the chloroplast genome and (2) the use of a heterologous probe for hybridization.

5. To wash off the unhybridized probe, cut a corner of bag pour off hybridization solution, open bag up, and float filters in a Pyrex dish at 50°C in 2× SSC,

   0.5% sarkosyl four times for 20 minutes each time. If possible, use a shaking water bath or a shaker-incubator. Dry the filters on a paper towel at room temperature, mark the filters with 32P-labeled ink. Put Saran wrap over the filters so they don't stick to the film. Set up an autoradiogram with X-ray-5 film and expose in the -70°C freezer for 3 days.

REFERENCES

1. Herrmann, R.G. (1982). In "Methods in Chloroplast Molecular Biology". (M. Edelman, R.B. Hallick and N.H. chua, eds.), 259-280. Elsevier, Amsterdam.
2. Kolodner, R., and Tewari, K.K. (1979). Proc. Natl. Acad. Sci. U.S.A. 76. 41-45.
3. Krebbers, E.T., Larrinua, I.M., McIntosh, L., and Bogorad, L. (1982). Nucleic Acids Res. 10,4985-5002.
4. McIntosh, L., Poulson. C., and Bogorad, L. (1980). Nature (London) 228, 556-560.
5. Ohyama, K., Fukuzawa, H., Kohchi, T., Shirai, H., Sano. T., Sano. S. Umesono. K., Shiki. Y., Takeuchi, M., Chang, Z., Aota, S., Inokuchi, H.(1986). Nature (London) 322, 572-574.
6. Palmer, J.D. (1982). Nucleic Acids Res. 10,1593-1605.
7. Palmer, J.D. (1983). Nature (London) 301, 92-93.
8. Palmer, J.D. (1985). Ann. Rev. Genet. 19: 325-354.
9. Palmer, J.D., and Thompson. W.F. (1981), Proc. Natl. Acad. Sci. U.S.A. 78, 5533-5537.
10. Schleif, R.F., and Wensink, P.C. (1981). "Practical Methods in Molecular Biology." Springer-Verlag, Berlin and New York.
11. Shinozaki, K., Ohme, M., Tanaka, M., Wakasugi, T., Hayashida, N., Matsubayashi, T., Naita, N., Chunwongse, J., Obokata, J., Yamaguchi-Shinozaki, K., Ohto, C., Torazawa, K., Meng, B.Y., Sugita, M., Deno, H., Kamogashira, T., Yamada, K., Kusuda, J., Takaiwa, F., Kato., A., Tohdoh, N., Shimada,
12. H., and Sugiura, M.(1986). EMBO J. 5, 2043- 2049.
13. Southern, E.M. (1975). J. Mol. Biol. 98, 503-517.
14. Schuler, M.A. and Zielinsky, R.E. 1989. Methods in plant Molecular biology. Academic Press, Inc. New York.

39

DNA SEQUENCING

Obtaining the nucleotide sequence of isolated DNA/gene is an important step in determining their identity and analysing their character. Sequence information is pre-requisite for planning any substantial manipulation of the DNA. There are two main sequencing methods that are widely used in the laboratories; (i) The chemical degradation by Maxam and Gilbert (1977) method; and (ii) enzymatic, chain termination or dideoxy chain termination method developed by Sanger *et al.* (1977)

(i) DNA SEQUENCING BY THE MAXAM AND GILBERT METHOD

This method of DNA sequencing makes use of chemical reagents for bringing about base specific cleavage of the DNA. The DNA strand to be sequenced must be radiolabelled at one end with a 32P phosphate group. An altered base can be removed from the sugar phosphate backbone of DNA. The molecule is cleaved with piperdine at the sugar residue lacking the base. When end of the base specific reagents is used in a limited reaction with end labelled DNA, a nested set of the end labelled fragments of different lengths are generated. All five reaction may be performed with different reagents specific for G, G+A, T+C, C, A>C. These sequencing reactions are analysed by running side by side on a 6-20% polyacrylamide 7M urea sequencing gel at high voltage to minimise the DNA secondary structure. The labelled DNA bands are obtained after autoradiography on X-ray film and the sequence can be read.

(ii) DNA SEQUENCING BY DIDEOXY CHAIN TERMINATION METHOD

This method is now favoured for large scale sequence determination. In practice, the Klenow fragment of DNA polymerase 1 which lacks the 5'-3' exonuclease activity of the intact enzyme is used to synthesize a complementary copy of the single stranded DNA sequence. Initiation of DNA synthesis requires a primer and usually this is chemically synthesized oligonucleotide which is annealed close by DNA synthesis is carried out in the presence of four deoxynucleotide, one of which is labelled with ^{32}P, and in four separate incubation mixtures containing a low concentration of one each of the four dideoxynucleotide triphosphate analogues. Therefore, in each reaction there is a population of partially synthesized radioactive DNA. Molecules having a common 5' but each varying a length to a case specific 3' end. After a suitable incubation period of DNA in each mixture, is denatured electrophoreses side by side and the radioactive bands of single stranded DNA detected by autoradiography. Further any substantial DNA sequence undertaking will necessitate computer analysis of the data for example, a computer search of the sequence for all known restriction endonuclease target sites will provide a complete and precise restriction map.

A complete protocol of USB DNA sequencing kit is given in this chapter which is as follows:

1. Preparation of Sequencing Gel and casting

The following are recipes for typical sequencing gel solutions. There are many variations in current use, but follow the protocols mentioned in the DNA sequencing kit of USB (Sequence version 2.0) and LKB Pharmacia for macrophor sequencing system.

Buffers and Solutions

Gel casting assembly (Macrophor Sequencing System LKB-Pharmacia): Polyacrylamide mix (95g Acrylamide, 5g Bisacrylamide, dissolved in 140 ml water and make the volume upto 250 ml with distilled water); Urea; TBE buffer; 10% Ammonium persulphate; TEMED; 10% Acetic acid.

Gel recipes (for 100ml of gel solution)

	Polyacrylamide gel concentration (%)						
	3	4	5	6	8	12	16
Urea (g)	42	42	42	42	42	42	42
Acrylamide/ Bisacrylamide T40C5 (ml)	7.5	10	12.5	15	20	30	40

10× TBE (ml)	10	10	10	10	10	10	10
ddH_2O (ml)	47.5	45	43	40.5	35	25	15
10% APS (ml)	0.8	0.8	0.8	0.8	0.8	0.8	0.8
TEMED (µl)	87	80	80	80	80	70	47

Dissolve urea in sterile double distilled water (sometimes incubation at 37°C is required). Add polyacrylamide mix and TBE. Adjust volume to 100ml with ddH_2O, filter and degas. When ready to pour add freshly prepared 10% ammonium persulfate and TEMED.

Protocol

1. Clean the glass plates with detergent, rinse throughly with water and air dry. Rinse with ethanol and air dry. Treat thermostatic plate with repel-Silane and notched glass plate with bind-Silane, rinse with distilled water, then with ethanol and air dry.
2. Arrange the glass plates for assembly with the silane-treated sides facing inward and with the bottom edges slightly offset. Insert 0.2 mm (top) and 0.4 mm (bottom) wedge spacers between the long edges of the plates. Alternatively, use 0.4 mm spacers for uniform gels. If the plates are to be clamped together, place a thin bead of petroleum jelly between the plates and the spacers before clamping.
3. To initiate gel formation, add 800 µl of 10% ammonium persulphate and appropriate amount of TEMED to the mixture and swirl to mix.
4. To pour the gel solution, tilt the Macro Mould to the middle portion. Apply solution just in front of notched glass plate while sliding it up the slope. Apply a slight pressure on the glass plate all the time while sliding.
5. Insert the sample application comb at the top of the plates, or the surface-former if a shark-tooth comb is to be used. Fix the comb clamps.
6. After pouring the gel, lay the plates in a horizontal position for 15 min to allow the gel to polymerise.

2. Sequencing Reaction

All sequencing reactions are run in EP tubes (0.5ml). These should be kept capped to minimize evaporation of the small volumes employed. Additions should be made with micropipetes using disposable-tips and care should be taken not to contaminate stock solutions. The solutions must be thoroughly mixed after each addition, typically by 'pumping' the solution two or three times with the micropipette, avoiding the creation of air bubbles. At any stage where the possibility exists for some solution to cling to the walls of tube, it should be centrifuged. These reactions can be completed in 1 h. with proper care and experience.

Buffers and solutions

Sequencing kit (USB); Ethanol; γ-S^{35} dATP; 2 M NaOH; 3 M Sodium acetate, 2 mM EDTA.

Protocol

1. Denaturation of template DNA: Use dsDNA directly for sequencing. Alkaline denaturation of template gives good results. The DNA is denatured by adding 0.1 volume of 2 M NaOH, 2 mM EDTA and incubating for 30 min at 37°C. The mixture is neutralized by adding 0.1 volume of 3M sodium acetate (pH 5.2) and DNA precipitated with 2-4 volumes of ethanol (-70°C, 15min). After washing the pelleted DNA with 70% ethanol, it is re-dissolved in 7 µl of distilled water.
2. Annealing of template and primer:
   In a EP tube (0.5ml) combine the following:

Primers (Forward or Reverse)	1 µl
Reaction Buffer	2 µl
DNA	7 µl (5 µg alkali denatured plasmid DNA)

   The total volume should be 10 µl. Anneal by heating for 2 min at 65°C, then cool slowly to < 35°C over 15-30 min. centrifuge briefly and chill on ice for use in labelling reaction (see 6).
3. While cooling, label, fill and cap tubes with 2.5 µl of each Termination Mixture (G, A, T and C). Use mixtures from RED-CAPPED tubes for dGTP. Keep covered at room temperature.
4. Dilute Labelling Mix 5-fold to working concentration if needed, dGTP (GREEN). Retain for use in step 6b.

Labeling Mix	-µl (Typically 2µl)
H_2O	-µl (Typically 8µl)

5. Prewarm each of the 4 termination tubes from step 3 (G, A, T & C) at 37°C in water bath.

6. Labelling reaction
   (a) Dilute the Sequence Version 2.0 enzyme 1:8 in ice-cold Enzyme Dilution Buffer. Only enough enzyme for immediate use should be diluted (2μl per reaction); diluted enzyme should be stored on ice for no more than 60 minutes. NEVER dilute Sequence DNA Polymerase in labelling mix, Dithiothreitol (DTT) solution or other non buffered solutions. Enzyme may be added to premixed cocktails only after all other materials including Enzyme Dilution Buffer are added.
   (b) To ice-cold Annealed DNA Mixture (10 μl)

   Add:

DTT (0.1 M)	1 μl
Diuted Labelling Mix	2 μl
γ-^{35}S dATP	0.5 μl
Diluted Sequences Polymerase	2 μl
Total	15.5 μl

   Mix and incubate at room temperature for 2-5 min.
7. Termination Reaction

   Transfer 3.5 μl of labelling reaction to each termination tube (G, A, T & C) mix and continue incubation reaction at 37°C for 5 min.
8. Stop the reaction by adding 4 μl of stop solution.
9. Heat the samples to 75-80°C for 2 min and load immediately on the pre run gel. Use 2-3 μl in each lane.

3. Sequencing Gel Pre Run

The quality of the gel electrophoresis is often the factor which limits the extent of sequence information that can be determined in a single sequencing experiment. The length of time the gel is run will determine the region of sequence that is readable. Under optimal conditions, 300 or more base can be read starting at the bottom of a gel. Unfortunately, many factors can reduce this resolution. Among these are the quality of reagents used, the polymerization, the temperature of the gel during electrophoresis and proper drying of the gel after running. In short, the greatest care should be given to the pouring and running of sequencing gels.

Buffers and Solutions

TBE buffer; Sequencing apparatus (LKB-Pharmacia).

Protocol

1. Carefully remove the comb. Immediately rinse the sample wells or surface with sterile double distilled water to remove unpolymerised acrylamide.
2. Place the gel in its associated electrophoresis apparatus and add lx TBE, making sure that the sample wells are filled with buffer and are free of air bubbles.
3. Prerun the gel at 1200-1500 V for 30-60 min.

4. Sequencing gel electrophoresis

Buffers and Reagents

TBE buffer; Sequencing apparatus (LKB-Pharmacia).

Protocol

1. After the gel has been pre-electrophoresed and the samples are denatured, switch off the power supply and wash the wells with buffer. (Note: If this wash is omitted, loading can be difficult and instead of bands smear may appear). Load 2-3 μl aliquots of denatured sequencing reactions (from step 9 of sequencing reaction) into adjacent wells, in sets of four. If you will be running multiple samples from each set of sequencing reactions, remember to allocate four lanes of the gel for each loading.
2. Reconnect the gel to the power supply and electrophoreses at 3500 V (150mA, 50W). If you plan to load additional samples of the same sequencing reactions continue electrophoresis until the xylene cyanol in the most recently loaded set is 4-5 cm from the bottom of the gel (about 2 h for a 40 cm gel). At this point, switch off the power, load the next aliquot of the samples into an adjacent group of four empty lanes, and continue electrophoresis as before.
3. Stop the final period of electrophoresis when the bromophenol blue in the samples loaded last reaches the bottom of the gel.

Sequencing Gel autoradiography

Buffers and Solutions

Whatman 3MM filter paper; Food wrap; X-ray film; X-ray cassette; Glacial acetic acid

Protocol

1. Switch off the power, and remove the gel/plate assembly from the electrophoresis apparatus. Lay the assembly in a horizontal

position with the notched plate facing upward. Carefully separate the plates so that the gel remains attached to the notched plate.

2. Place the notched glass plate with gel still attached to it in an exposure box containing approximately 21 of 10% acetic acid.
3. Agitate for 15 min.
4. To dry the gel, transfer carefully to a supporting sheet of 3 MM Whatman filter paper; cover with high-quality food wrap, and dry using a vacuum gel dryer.
5. In a darkroom, place the covered gel/support in a film cassette with an appropriately sized sheet of X-ray film next to the gel; expose for 4-16 hours at room temperature, then develop the film according to the manufacture's instructions.

REFERENCES

1. Rubin, C.M. and Schmix, C.W. 1980. Nucleic Acids Research, 8: 4613-4619.
2. Saluz. H.P. and Jost, J.P. 1986. Gene, 42: 151-157.
3. Saluz, H.P. and Jost, J.P. 1987. A laboratory Guide to Genomic Sequencing. A direct sequencing of Native uncloned DNA. Birkhauser, Basel. Boston.
4. Southern, E. 1975. J. Molecular Biol. 98: 503-517.
5. Maxam, A.M. and Gilbert, W. 1977. Proc. Nat. Acad. Sci. USA, 74: 560-564.
6. Sanger, F., Nicklens, S. and Coulson. A.R. 1977. Proc. Nat. Acad. Sci. USA, 74: 5463-5467.
7. Anon, 1996, Training manual on Molecular Biology and Crop biotechnology. CCS University, Meerut (India).

40

IDENTIFICATION OF PROTEINS BY TWO-DIMENSIONAL POLYACRYLAMIDE GEL ELECTROPHORESIS (PAGE) BASED ANALYSIS AND WESTERN BLOTTING

Two-dimensional gel electrophoresis is the most powerful protein separation technique and is applicable to soluble as well as membrane proteins. By combining iso-focusing with sodium dodecylsulphate-polyacrylamide gel electrophoresis (SDS-PAGE), a high resolution and a high separation capacity is obtained. These properties make two-dimensional polyacrylamide gel electrophoresis (2D-PAGE) the method of choice for the analysis of very complex protein mixtures. More than a thousand radioactively labelled proteins can be resolved and detected in a single two-dimensional gel (O'Farrell, 1975; Garrels, 1979).

Owing to the development of protein blotting onto chemical-resistant membranes such as coated glass-fibre and polyvinylidene difluoride (PVDF) membranes, many new proteins purified by 2D-PAGE were identified (Bauw *et. al.*, 1987: Celis *et. al.*, 1990). In this way, two-dimensional gel electrophoresis evolved from a purely analytical tool to a tool allowing the identification of the visualized protein.

1. TWO-DIMENSIONAL ELECTROPHORESIS OF PLANT PROTEINS

Two-dimensional gel electrophoresis, as described here, is identical to the method originally described by O'Farrell (1975). The electrophoresis apparatus for iso-focusing the SDS-PAGE are based on those described by O'Farrell (1975). The result of the 2D-PAGE is strongly influenced by the preparation of the proteins. Many problems encountered with 2D-PAGE, particularly when analysing plant proteins, are related to the isolation of the protein sample. Plant tissue generally has a lower protein content, and the presence of phenolic compounds, tannins, lipids, and others, interfere with the clean separation of the proteins. Most of these components bind covalently and/or non-covalently to proteins, altering their charge and behaviour.

Many different extraction methods for plant tissue proteins have been described (Damerval et. al. 1991). A good pretreatment of proteins for 2D-gel electrophoresis includes four steps: tissue homogenization, protein extraction (separation from other components), concentration, and solubilization of the protein. All steps are important and will influence the two-dimensional separation. Inefficient grinding of the tissue will lead to loss of protein and to an unreliable protein pattern. The presence of phenolics, lignins and tannins in the protein sample may result in protein modification, insolubility or streaking in the gel. Proteins not completely solubilized will not, or only partially, be found in the resulting two-dimensional gel protein pattern, and consequently, this pattern will not reflect the real protein content of the analysed tissue.

2. PROTEIN EXTRACTION WITH PHENOL

The phenol extraction procedure is adapted from Hurkman and Tanaka (1986). The method has been successfully used for 2D-PAGE of xylem tissue of poplar and various tissues of *Arabidopsisand* tobacco. Either freshly harvested or frozen tissue (-70°C) is used for the extraction. If the harvested tissue is stored at -70°C, do not grind the tissue before storage because grinding liberates the proteases, which will be enzymatically active upon thawing. To reduce the proteolytic activity of proteases, always keep the extract on ice. It is also possible to add a cocktail of protease inhibitors to the extract solution.

Buffers and solutions

Protein extraction buffer: 0.7 M. sucrose. 0.5 M Tris, 30 mM HCl, 0.1 M KCl and 1% z- mercaptoethanol.

Protocol

1. Weigh the tissue (0.5-1 g) and place in a chilled mortar. In the mortar, freeze the tissue with liquid N_2 and immediately grind with a pestle to a fine powder.
2. Add extraction buffer (3 ml) to the powder. Homogenise further to fine suspension and transfer to a Corex tube. Wash the mortar

twice with 1 ml of extraction buffer and transfer this solution to the same Corex tube. Vortex the extract solution thoroughly and keep it at 4°C for at least 30 min.

3. Centrifuge the suspension at 4000 g for 10 min in a swinging bucket rotor. Transfer the supernatant to a new Corex tube and keep at 4°C.
4. Resuspend the pellet in 3 to 5 ml of extraction buffer. The pellet can be further homogenized using a glass Dounce homogenizer. Vortex the suspension throughly and keep it at 4°C for 30 min.
5. Centrifuge the suspension again at 4000 g for 10 min. Add the supernatant to the first one.
6. To the combined supernatant, add 5 ml of water-saturated phenol. Mix the phenol and the extract thoroughly and keep on ice for 1h.
7. Separate the phases by centrifugation in a swinging bucket at 8000 g for 10 min. Transfer the upper phenol phase to a new corex tube. Add 3 ml of fresh water-saturated phenol to the water phase and the interphase, and repeat the extraction.
8. Combine both phenol phases. Precipitate the proteins from the phenol by addition of 20 ml 0.1 M ammonium acetate dissolved in methanol and place overnight at -20°C.
9. Obtain a protein pellet by centrifugation in a swinging bucket (maximum 4000 g) for 10 min. Pour off the supernatant and dissolve the pellet in 1ml of water.
10. Add nine volumes of cold acetone and place the solution at -20°C for at least 4 h.
11. Isolate the precipitated protein again by a centrifugation at a maximum of 4000 g in a swinging bucket. Pour off the supernatant and briefly dry the pellet in the air until the smell of acetone has disappeared.
12. Dissolve the protein pellet in the appropriate amount of protein sample buffer and freeze at -20°C.

3. ISOFOCUSING, THE FIRST DIMENSION

Initially, the electrical field is gradually increased to give the proteins enough time to penetrate the isofocusing gel and to avoid a massive precipitation of the protein. Partial precipitation, especially when a high protein concentration is used, is difficult to avoid, but most of it will redissolve during the isofocusing.

Buffers and solutions

1. 30% acrylamide (2x crystallized)
2. 2% bisacrylamide
3. 10% Nonidet-P40(NP40)
4. 10% ammonium persulphate (APS).
5. Protein sample buffer : 9.5 M urea, 2% w/v NP 40, 2% ampholines (pH 3.5-10), 100mM dithiothreitol (DTT).
6. Overlay buffer: 6M urea, 2% NP 40, 1% ampholines (pH 3.5-10), 100mM DTT. Composition of isofocusing gels: 5.5 g urea, 1.3 ml 30% acrylamide, 0.5ml 2% bisacrylamide, 2ml 10% NP 40, 1.5ml H_2O and 0.6ml ampholine mix (ampholines pH 3.5-10 : pH 4-7 : pH 5-8 in 4:2:2 proportion).

Protocol for First Dimension, Isofocusing

1. Perform the isofocusing in cylindrical rod gels. The gels are made in glass tubes 15 cm long with an inner diameter of 1.4mm. To clean the glass tubes after usage, follow the procedure : 24 h in 2M NaOH, rinse with H_2O, 24 h in 1M acetic acid, rinse with H_2O, rinse with methanol and dry.
2. Seal the bottom of the glass tubes with Parafilm. Place a mark 13cm from the bottom of the tube to indicate to where the tube must be filled with the gel solution.
3. Prepare the isofocusing gel solution in a flask using the composition described above. Dissolve the urea by slightly warming the solution (maximum 30°C).
4. Start the polymerization by the addition of 15 μl 10% APS and 10 μl N, N, N^1, N^1-tetramethylethylenediamine (TEMED). Then immediately transfer the gel solution into the glass tubes using a syringe with a long, narrow needle. Cautiously bring the needle to the bottom of the glass tube and gently extrude the gel solution from the syringe. No air bubbles should be trapped in the gel solution. Overlay the gel solution with 50 μl H_2O.
5. For a complete polymerization, leave the gels for at least 5 h at room temperature.
6. Cautiously remove the Parafilm from each glass tube and mount the tubes into the vertical isofocusing apparatus.
7. Rinse the bottom of the isofocusing gel with H_2O and remove potential air bubbles. Immerse the gel bottom in the lower electrode buffer which consists of 10mM H_3PO_4.

8. Remove the H_2O covering the top of the isofocusing gel and rinse the glass tube above the gel twice with H_2O without disturbing the gel surface. Carefully take away all remaining H_2O from the top of the isofocusing gel with a Hamilton syringe.
9. Form three separate layers, consecutively 20 µl protein sample buffer, 20 µl overlay buffer and 20 mM NaOH, on top of each other on the isofocusing gel. Fill the upper electrode chamber with the electrode buffer (20mM NaOH). The H_3PO_4 -containing chamber is connected to the positive pole and the NaOH one to the negative pole of a power supply. Perform the prerun by applying, consecutively, 200V for 15 min, 300 V for 30 min, and 400 V for 1h.
10. After this pre-run, disconnect the isofocusing apparatus from the power supply and remove the NaOH buffer from the upper chamber. With a syringe remove all liquid from the top of the gel tubes and wash the top twice with H_2O. Carefully take off all remaining H_2O.
11. Thaw protein samples at room temperature, warming the samples if necessary, but not above 35°C. Apply up to 150µl of the protein sample on top of the isofocusing gel. Overlay the protein sample with 20 µl of overlay buffer and fill the remaining part of the glass tube with fresh NaOH-electrode buffer. Fill the upper chamber with fresh NaOH solution which has been degassed and reconnect the isofocusing apparatus to the power supply.
12. Start the isofocusing at 100V and increase every 15 min by 100V until 400 V is reached. Run the gels at 400V for 18h.
13. After isofocusing, remove the gel tubes from the apparatus. Either perform the second dimension immediately or store the gels in their tube temporarily (maximum 2-3 days) at -20°C.

4. SECOND DIMENSION: SDS-PAGE AND PROTEIN STAINING

As acrylamide and bisacrylamide are toxic compounds, care should be taken when using them, especially when handling the solid form. Only the acrylamide solution is filtered through a 0.4 µm filter. All reagents are-stored at 4°C. Because the isofocusing gel is made of a very low acrylamide concentration, it stretches easily. Apply the rod gel on the second dimension without elongating it too much. Always wear gloves when handling the two-dimensional polyacrylamide gel to avoid fingerprints on the gel. Coomassie staining is less sensitive than silver staining, but it is the only method which gives a good indication of the quantity of each detected protein. The quantity of a protein spot is best estimated by comparing its intensity with that of a known amount of a reference protein run simultaneously on the gel.

Buffers and solutions

1. 30% acrylamide (2xcrystallized)
2. 2% bisacrylamide.
3. 2 M Tris-Cl pH 8.6.
4. 1 M Tris-Cl pH 6.8.
5. 20% SDS.
6. 10% APS.
7. Electrophoresis buffer: 60 g Tris, 144 g glycine, 10g SDS dissolved in 10 litre of water.
8. Second-dimension equilibration buffer: 60 mM Tris, 1% SDS, 20% glycerol, 50 mM DTT adjusted to pH 6.8 with HCl. Add a small amount of bromophenol blue. Composition of separation gels is give in Table 1.

Protocol for second dimension, SDS-PAGE

1. Always clean the glass plates (20 × 20cm) immediately after use with detergent, rinse extensively with water, and dry before storage. Before assembling, clean the side of the glass plates which will be in contact with the polycrylamide gel with ethanol.
2. Between the two glass plates, one front and one back plate place three 1.5- mm-thick plastic spacers in a U form. Place the spacers 0.5 cm from the border of the glass plates. Hold the formed cassette together with clamps. Completely seal the space between the spacers and the outer border of the plates with melted agarose.
3. Make the separation gel solution in a 100-ml vacuum flask by mixing the different components for the separation gels (see Table 1). Add SDS, APS, and TEMED after deaeration of the solution. Gently pour the solution between the glass plates to 2 cm from the top of the front plate and then cover the solution with a water layer. Pour the separation gel at least 5 h before use.

Table 1 : Composition of separation gels (Laemmli, 1970)

	Protein (molecular mass range in kDa				
	>70	100	70	50	40
Acrylamide concentration:	7.5%	10.0%	12.5%	15.0%	17.5%
Acrylamide	7.5	10.0	12.5	15.0	17.5
Bisacrylamide	3	3.96	5.1	6	7
Tris-Cl (pH 8.7)	11.25	11.25	11.25	11.25	11.25
H_2O	8	4.5	0.9		-
20% SDS	0.15	0.15	0.15	0.15	0.15
10% APS	0.1	0.1	0.1	0.1	0.1
TEMED	0.02	0.02	0.02	0.02	0.02

All volumes are expressed in ml. The total volume is sufficient to make one 1.5-mm thick gel.

4. Make the stacking solution by mixing 1.7 ml 30% acrylamide, 0.6ml 2% bisacrylamide, 1.25 ml Tris-Cl (pH 6.8), and 5.7 ml H_2O. After deaeration of this solution, add 0.05 ml 20% SDS, 0.05ml 10% APS, and 0.005ml TEMED and place 1.5 cm of stacking solution on top of the gel. Overlay the stacking gel with H_2O-saturated isobutanol.
5. Before mounting the polyacrylamide cassette in the electrophoesis apparatus, rinse the top of the stacking gel extensively with H_2O to remove all isobutanol. Remove the water from the top of the gel, and mount the cassette in the electrophoresis apparatus.
6. Thaw the glass tubes with the isofocusing gels at room temperature. Connect a water-filled 5ml syringe with the basic side of the glass tube. Gently extrude the rod gel from the tube onto a plastic sheet by slowly exerting pressure on the syringe.
7. Place the sheet on the border of the front glass plate and gently push the isofocusing gel against the back plate Cautiously remove the plastic sheet, keeping the isofocusing gel between the two glass plates. Starting from one side, going gradually to the other, gently press the rod gel between the two plates with a round-shaped spatula, until the isofocusing gel makes contact with the stacking gel over its total length. When pressing the isofocusing gel onto the stacking gel, air bubbles between the gels are

systematically removed. To equilibrate the isofocusing gels, apply 1 ml of second-dimension equilibratin buffer on the gel.

8. After 10 min of equilibration, fill the upper chamber with electrophoresis buffer. Start the electrophoresis at 30V. When the migration front (bromophenol blue serves as indicator) has entered the separation gel, the voltage can be increased. Run the gels overnight at a constant voltage of 70V. Run the electrophoresis until the migration front has left the bottom of the gel.

5. COOMASSIE STAINING OF THE GEL

Buffers and solutions

1. Coomassie staining solution: 2 g/l Coomassie Brilliant Blue R 250, 45% methanol, 7% acetic acid.
2. Destaining solution: 10% methanol, 10% acetic acid.

Protocol

1. Remove the polyacrylamide gel from between the glass plates and immerse in 150 ml of Coomassie staining solution. Place on a rotary shaker, and gently shake the gel overnight.
2. Remove the staining solution and rinse the gel extensively with H_2O. Place the gel overnight in the destaining solution on a rotary shaker.

6. PROTEIN BLOTTING ONTO POLYVINYLIDENE DIFLUORIDE MEMBRANES

Immobilon is a very hydrophobic membrane and it is necessary to first use methanol to wet it. The two different sides of the membrance, easily distinguished when wet, have different properties: one side is hydrophilic-water remains equally dispersed; the other is hydrophobic-water tends to form drops. To reduce the occurrence of air bubbles, always place the hydrophilic side of the PVDF membrane facing the polyacrylamide gel. PVDF blots can be stored for years at 4°C after drying in the air but the colour of the spots can fade during storage. At any time, these PVDF blots can be restained with Amido Black after wetting the membrane with methanol.

A complete transfer of proteins from the polyacrylamide gel is only obtained if the gel is blotted immediately. It is possible to transfer Coomassie-stained gels, but the efficiency of transfer is much lower. To blot Commassie-stained gels, first wash them extensively with distilled water until all acid and methanol is removed. The proteins are remobilized by equilibrating the gel for 2 h in the pre-equilibration buffer. The electrotransfer is then performed as described.

7. PROTEIN BLOTTING ONTO IMMOBILON

Buffers and solutions

1. Pre-equilibration buffer: 50 mM H_3BO_3, 0.1% SDS adjusted to pH 8 with NaOH.
2. Transfer buffer: 50 mM Tris, 50 mM H_3BO_3.
3. Amido Black staining solution: 1 g/l Amido Black, 40% methanol, 7% acetic acid.

Protocol

1. After the electrophoresis, place the polyacrylamide gel immediately in 150 ml of preequilibration buffer. Allow the gel to swell in this buffer for 1.5 h on a rotary shaker.
2. Meanwhile cut Immobile to the size of the polyacrylamide gel. First, immerse the PVDF membrane for 5 min in methanol and then wash thoroughly with distilled water to remove all methanol. Place the membrane in water for at least 1 h on rotary shaker with gentle agitation.
3. To avoid air being trapped between the different layers of the blotting sandwich, assemble the sandwich in a box filled with transfer buffer and pre-wet each part. Make the blotting sandwich by mounting the following components on top of each other: a porous pad, four sheets of Whatman 3MM paper, the equilibrated polyacrylamide gel, the PVDF membrane, four sheets of Whatman paper, and a porous pad. Make sure no air is trapped between two layers.
4. Mount the blotting sandwich into the gel holder unit and insert into the transfer cell filled with the transfer buffer. Insert the blotting sandwich into the transfer cell so that the polyacrylamide gel is faced towards the negative electrode and the blotting membrane towards the positive.
5. Carry out the electrotransfer at a constant voltage of 35 V and the transfer will be complete after 8 h. The transfer may also be performed overnight at 30V.
6. After transfer, disassemble the blotting sandwich and place the PVDF membrane in distilled water and gently agitate for 5 min. Then transfer the membrane into the Amido Black staining

solution. After 10 min of gentle agitation in the staining solution, pour off the solution and rinse the membrane with water until the proteins are visible as blue spots on a white background.

8. SEMI-PREPARATIVE TWO-DIMENSIONAL GEL ELECTROPHORESIS

The maximum amount of protein which can be applied onto a two-dimensional gel depends on many factors: the size of the polycarylamide gel, the purpose of the separation (analytical versus preparative), the protein detection technique used, and the composition of the protein mixture. The amount of plant material necessary to generate a two-dimensional gel is largely dependent on the tissue. A good quality Coomassie-stained, two-dimensional gel can be generated from 0.1 g leaf or 0.2 g root tissue from *Arabidopsis*, wheareas, lg of xylem tissue of poplar is needed for a comparable two-dimensional gel. On these two-dimensional gels, 100 to 200 μg of protein can be loaded. This estimation is based on extrapolation from the amounts measured by sequence analysis of different protein spots.

The highest possible amount of protein is always applied on the two-dimensional gels which will be used for further sequence analysis of the proteins. After electrophoresis, these gels are transferred to PVDF membranes. This blotting step is also a concentration step: the protein is condensed on a surface. The amount of protein applied on semi-preparative gels must not be too large, otherwise the protein spot (s) of interest will overlap with neighbouring proteins after blotting.

10. PROTEIN IDENTIFICATION OF 2D-PAGE-SEPARATED PROTEINS BY SEQUENCE ANALYSIS

A first step to sequence analysis is cutting out the protein spots of interest and collecting them from multiple gels until the necessary amount is reached. Only blots on which the protein pattern is completely recognized, the spots of interest are well resolved from the others, and the spots are clearly visible with Amido Black (spot representing more than 0.1 μg), should be used. When the sample will be subjected to an *in situ* digestion with a protease, only the membrane surfaces on which the proteins are bound should be used.

The availability of amino acid sequences of a protein of interest allows a straightforward isolation of corresponding cDNAs or genes (Bauw and Montagu, 1997). Corresponding degenerate oligonucleotides are synthesized from the most suitable amino acid sequences. Regions with less degeneration (rich in one or two codon-encoded amino acids) are chosen especially, whereas, regions containing *Leu*, *Ser* and *Arg* (encoded by six codons) are avoided. These synthetic oligonucleotides are used in two approaches; first, they are used for direct screening of cDNA libraries; and second, two oligonucleotides, one in sense and one in antisense orientation, are used as templates in a polymerase chain reaction (PCR) with first-strand cDNA or genomic DNA. The isolated PCR fragment is then used to screen cDNA or genomic libraries.

10. *IN SITU* TRYPSIN DIGESTION ON PVDF

Buffers and solutions

1. Digestion buffer: 0.1 M Tris-Cl pH 8.5.
2. 0.1 M $CaCl_2$.
3. 5% reduced Triton X-100.

Protocol

1. Cut out the protein spots of interest from the blots with a scalpel and collect in an Eppendorf tube. Only cut out the membrane area to which the protein is bound.
2. If necessary trim the membrane area to pieces of 2 × 2 mm. Add methanol (100 μl) to the Eppendorf tube and after 5 min discard the methanol solution. To remove all remaining methanol, wash the membrane pieces five times with 200 μl of distilled water.
3. To the membrane pieces, add 100 μl of digestion buffer and 10 μl of 5% reduced Triton X-100 and leave for 2h at room temperature.
4. Make a fresh solution of 1 mg/ml porcine trypsin in digestion buffer. To each Eppendorf tube, add 5 μl of 0.1M $CaCl_2$ and 1 or 2 μl of the trypsin solution. Incubate the digestion at 37°C for at least 6 h.
5. After digestion, transfer the digestion solution to a new Eppendorf tube. Wash the membrane pieces successively with 200 μl 6% trichloroacetic acid (TFA) and 200 μl H_2O. Add these wash solutions to the digestion solution. Store the peptide solution at-20°C until HPLC analysis.
6. Separate the peptides by reversed-phase HPLC on a C4 column. Equilibrate the column with solvent A (0.1% TFA) and maintain for 5 min after injection of the peptide solution. Elute peptides in a linear gradient to 55% solvent B (70% acetonitrile, 0.1% TFA) for 50 min. The column eluate is monitored by UV absorption at

214 nm and the peptides are collected manually in an Eppendorf tube.

11. WESTERN BLOTTING

This technique is used to detect proteins of a particular specificity. Particularly when a transferred gene expresses in transformed cells, the translated product in the form of proteins can be identified by this technique, the extracted proteins are subjected to polyacrylamide gel electrophoresis (PAGE) and are then transferred onto nitrocellulose membrane, to which they bind. Nitrocellulose membrane is then used for probing with a specific labelled antibody (antibody will not hybridize with protein, but bind with it,). The antibody may be lebelled with ^{125}I and the signal is detected again with autoradiography.

Protocols

1. Prepare the following solutions:
   * running buffer: 3.03 g Tris base; 14.4 g glycine; 200 ml methanol; distilled water to 1 litre.
   * low-salt blocking buffer: 20 mM Tris-HCl, pH 7.6; 0.9% (w/v) sodium chloride; 0.05% (v/v) Triton X-100; 0.5% casein.
   * low-salt wash buffer: 20 mM Tris-HCl, pH 7.6; 0.9% (w/v) sodium chloride; 0.05% Triton X-100.
2. Measure the length and breadth of the gel and cut one piece of nitrocellulose and two pieces of 2 MM filter paper to the size of the gel.
3. Soak the Scotch-Brite pads or sponge pads of the apparatus in running buffer.
4. Assemble a 'sandwich' comprising 3 MM filter paper, gel, nitrocellulose filter, and 3MM filter paper. Ensure that all air-bubbles between gel and filter are removed.
5. Insert the 'sandwich' into the blot apparatus, ensuring that the nitrocellulose filter is between the gel and the anode (+charge). SDS-protein complexes are negatively charged and will move to the anode.
6. Electroblot for either 60V, 0.22 A for 3h, or 30V. 0.1 A overnight.
7. Remove the filter and wash in two changes of low-salt blocking buffer, during a 30 min period. This blocks non-specific protein-binding sites on the filter.

8. Place the filter on a glass plate in a humidity chamber (freezer box containing wet tissues). Spread 1 ml of low-salt buffer containing a 1/100 dilution of antiserum on to the side of the filter on to which the proteins were transferred. Incubate for 1h at room temperature.
9. Wash the filter in four changes of low-salt buffer during a 1h period, with gentle shaking.
10. Detect antibodies bound to protein on the filter using either anti-1g conjugated to enzyme, e.g. horseradish peroxidase, or ^{125}I-labelled protein A. develop autoradiograms.

REFERENCES

1. Bauw, G., De Loose, M., Inze, D., Van Montagu, M. and Vandererckohove, J. 1987. Proc. Natl. Acad, Sci, USA, 84 : 4806-4810.
2. Bauw, G. and Montagu, M.V. 1997. In: Differentially Expressed genes in plants-A Bench Manual (Hansen, E. and Harper, G. eds.). Taylor and Francis Ltd., John Street, London, pp-105-118.
3. Celis, J.E., Cruger, D., Kill, J., Dejgaard, K., Lauridsen, J.B., Ratz., G.P., Basse, B., Celis, A., Rasmussen, H.H., Bauw, G. and Vandekerckhove, J. 1990. Electrophoresis, 11, 242-254.
4. Damerval, C., Zivy, M., Granier, F. AND De Vienne, D., 1991 In: Advances in electrophoresis.
5. Garrels, J.I., 1979. Journal of Biological Chemistry, 254, 7961-7977.
6. Hurkman, W.J. and Tanaka, C.K., 1986. Plant Physiology 81: 802-806.
7. Laemmli, U.K.1970. Nature, 227, 680-685.
8. O'Farrell, P.H. 1975. Journal of Biological Chemistry, 250, 4007-4021.

Appendix 1

Table : Components and their concentrations used in different plant tissue culture media

Constituent	Concentration in culture medium (mg/litre)							
	MS	**SH**	**B5**	**GD**	**KM**	**LM**	**NN**	**SI**
$Ca(NO_3)_2$			241					
KNO_3	1900	2500	2500	1000	1900	1900	950	-
$NH_4.NO_3$	1650			1000	600	1650	720	
$NH_4.H_2PO_4$		300						
$(NH_4)_2.SO_4$			134					
$MgSO_4.7H_2O$	370	400	250	35	300	1850	185	370
$CaCl_2.2H_2O$ 440	200	150		600	22	220	440	
KCl				65	300			1400
KH_2PO_4	170			300	170	340	68	170
$NAH_2PO_4\ H_2O$			150					
$MnSO_4\ H_2O$	10.0	10.0	1.12	10.0	21			
$MnSO_4\ 4H_2O$	22.3						25	22.3
K1	0.83	1.0	0.75	0.8	0.75	4.15		0.83
H_3BO_3	6.2	5.0	3.0	0.3	3.00	31	10	6.2
$ZnSO_4\ 7H_2O$	8.6	1.0	2.0	0.3	2.00	43	10	8.6
$CuSO_4\ 5H_2O$	0.025	0.2	0.025	0.025	0.025	0.50	0.025	0.025
$Na_2MoO_4\ 2H_2O$	0.25	0.1	0.25	0.025	0.25	1.25	0.25	0.25
$CoC1_2\ 6H_2O$	0.025	0.1	0.025	0.025	0.025	0.125		0.025
$FeSO_4\ 7H_2O$ 27.8	15.0	27.8	27.8		27.8	27.8	27.8	
Sequestrene 330 Fe					28			
Na_2 EDTA	37.3	20.0	37.3	37.3		37.3	37.3	37.3
Nicotinic acid	0.5	5.0	1.0	0.1		0.5	5	0.5
Pyridoxine-HCl	0.5	0.5	1.0	0.1	1	0.1	0.5	0.5
Thiamine-HCl	0.1	5.0	10.0	1.0	1	0.1	0.5	0.1
D-Biotin				0.2	0.01		0.05	
Folic acid					0.4		0.5	
myo-Inositol	100	1000	100	10	100	100	100	100
Glycine	2.0			4	0.1		2	2.0
Glutamine								1462
Sucrose	30000	30000	20000	20000	20000	30000	20000	10000

Additional components of KM medium (mg/litre): nicotinamide 1; calcium D-pantothenate 1; folic acid 0.4; p-aminobenzoic acid 0.02; choline chloride 1; riboflavin 0.2; ascorbic acid 2; vitamin A 0.1; vitamin B12 0.02; sodium pyruvate 20; citric acid 40; malic acid 40; fumaric acid 40; fructose, ribose, xylose, mannose, rhamnose, cellobiose, sorbitol, and mannitol, 250 each; adenine 10: guanine, thymine, uracil, hypoxanthine, and xanthine, 0.03 each; all 21 L-amino acids 0.1 each except glutamine 5.6, alanine 0.6, glutamic acid 0.6 and cysteine 0.2.

MS= Murashige and Skoog (1962), SH= Schenk and Hildebrandt (1972), B5= Gamborg *et. al.* (1968) GD= Gresshoff and Doy (1974), KM= Kao and Michayluk (1975), LM= Litvay *et. al.* (1985) NN= Nitsch and Nitsch (1969) and Sl= Franklin *et al.* (1991)

Source : Dixon. R.A. and Gonzales. R.A. (eds). 1994. Plant cell culture- A practical approach. 2nd Edition, Oxford University press, Oxford. New York. Tokyo.

Appendix 2

Table : Commonly used plant growth regulators

Class	Name	Abbre-viation	Uses
Auxin	Indole-3-acetic acid	IAA	Use for callus induction at 10-30 μM. Lowering to 1-10 μM can stimulate organogenesis. It inactivated by light and readily oxidized by plant cells. The synthetic auxins below have largely superceded IAA for tissue culture studies.
	Indole-3-butyric acid	IBA	Use for rooting shoots regenerated via organogensis. Either maintain at low concentration (1-50 μM) throughout rooting process, or expose to high concentration (100-250 μM) for 2-10 days and then transfer to hormone-free medium. Can also use as a dip for *in vitro* or *ex vitro* rooting of shoots.
	2,4 dichlorophenoxya-cetic acid	2,4-D	Most commonly used synthetic auxin for inducing callus and maintaining callus and suspension cells in dedifferentiated state. Usually used as sole auxin source (1-15 μM), or in combination with NAA.
	p-Chlorophenoxyacetic acid	pCPA	Similar to 2,4-D, but less commonly used.
	1 - N aphthaleneacetic acid	NAA	Synthetic analogue of IAA. Commonly used, either as sole auxin source (2-20 μM) for callus induction and growth of callus and suspension cultures; 0.2-2μM for root induction), or, in combination with 2,4-D.
Cytokinin	6-Furfurylaminopurine (Kinetin)	K	Often included in culture media for callus induction, growth of callus and cell suspension, and induction of morphogenesis (1-20μM). Higher concentration (20-50μM) can be used to induce the rapid multiplication of shoots, axillary/adventitious buds, or meristems.
	6-Benzylaminopurine	BAP	Included in culture media for callus induction, and growth of callus and cell suspensions (0.5-5.0μM), and for induction of morphogenesis (1-10μM). More commonly used than kinetin for inducing rapid multiplication of shoots, buds, or meristems (5-50μM).

Class	Name	Abbre-viation	Uses
	N-Isopentenylamino-purine	2ip	Less commonly used than K or BAP for callus induction and growth (2-10μM), induction of morphogenesis (10-25μM), or multiplication of shoots, buds, or meristems (30-50μM).
	Zeatin	Zea	Seldom used in callus or suspension media. Can be used for induction of morphogenesis (0.05-10μM). Zea is thermolabile and must not be autoclaved.
Gibberellin	Gibberellin A_3	GA_3	Seldom used in callus or suspension medium (one exception being potato). Can promote shoot growth when added to shoot induction medium at 0.03-14 μM. also used to enhance development in embryo/ovule cultures (0.3-48 μM). GA_3 is thermolabile and must not be autoclaved.
Abscisic acid	Abscisic acid	ABA	Used at concentrations of 0.04-10 μM to prevent precocious germination, and promote normal development of somatic embryos (60).

Note: The concentrations given represent values taken from the literature for a range of plant species. Specific applications for species not described in the literature should be determined experimentally.

Source: Dixon, R.A. and Gonzales, R.A. (eds.) 1994. Plant cell culture - A Practical Approach, 2nd edition. Oxford University Press, Oxford, New York, Tokyo.

Appendix 3

Table : Table : Plant Growth Regulators, their Preparation and Storage

Growth Regulators	IBA	KIBA	IAA	2,4 D	4-CPA	IAA-L Asparti c Acid	IAA-L Ala-nine	IAA-Glyc-ine	IAA-L Phenyla-lanine	NAA	2IP	BA	Kinetin	Zeatin	Adenine Hemi-Sulate	Adenine
Preparation & Storage																
Solvent	EtOH/ IN NaOH	TC/ Water	EtOH/ IN NaOH	EtOH/ IN NaOH	EtOH	0.5N NaOH	0.5N NaOH	0.5N NaOH	0.5N NaOH	IN NaOH	IN NaOH	IN NaOH	IN NaOH	IN NaOH	TC/ Water	TC/ Water
Diluent	TC Water	–	TC/ Water	TC/ Water	TC/ Water	TC/ Water	TC/ Water	TC/ Water	TC/ Water	TC/ Water	TC/ Water	TC/ Water	TC/ Water	TC/ Water	NA	NA
Powder Storage	0.5 °C	0.5 °C	-0 °C	RT	RT	-0 °C	-0 °C	-0 °C	-0 °C	RT	-0 °C	RT	-0 °C	-0 °C	RT	RT
Liquid Storage	-0 °C	0.5 °C	-0 °C	0.5 C	0.5 C	-0 °C	-0 °C	-0 °C	-0 °C	0.5 °C	-0 °C	0.5 °C	-0 °C	-0 °C	0.5 °C	0.5 °C
Sterilization	CA/F	CA/F	CA/F	CA	CA	F	F	F	F	CA	CA/F	CA	CA/F	CA/F	CA	CA
Working conc. (mg/L)	0.1-10.0	0.1-10.0	0.1-3.0	0.01-5.0	1.0-10.0	0.01-5.0	0.01-5.0	0.01-5.0	0.01-5.0	0.1-10.0	1.0-30.0	0.1-5.0	0.1-5.0	0.01-5.0	50-250	50-250

Growth Regulators	BPA	DPU	KR	BAR	2iPR	ZR	GA3	ABA	t-Cinnamic	Succinic acid	Phloro flucinol	Triaconta	Colchicine
Preparation & Storage													
Solvent	IN NaOH	DMSO	IN NaOH	IN NaOH	TC Water	TC Water	EtOH	IN NaOH	EtOH/ Acetone	TC Water	TC Water	See note	TC Water
Diluent	TC Water	TC Media	TC Water	TC Water	NA	NA	TC Water	TC Water	TC Water	NA	NA	TC Media	NA
Powder Storage	-0 °C	RT	0.5 C	-0 °C	-0 °C	-0 °C	RT	0 °C	RT	0.5 °C	RT	0.5 °C	RT
Liquid Storage	-0 °C	0.5 C	0.5 C	-0 °C	-0 °C	-0 °C	0.5 °C	-0 °C	0.5-0 °C	0.5-0 °C	0.5 -0 °C	0.5 -0 °C	0.5 °C
Sterilization	CA/F	F	F	F	F	CAF	CA/F	CA/F	CA/F	CA/F	CA/F	CA	F
Working Conc. (mgL)	0.1-5.0	0.1-1.0	0.01-5.0	0.01-10.0	0.1-10.0	0.1-10.0	0.1-10.0	0.1-10.0	0.1-10.0	0.01-1.0			

Note : All vitamins can be dissolved in water and stored at -20°C

F = Filter sterilize

CA = Coautoclavable with other media compouents

CA/F = Coautoclavable with other media components, however some loss of activity many occur This can be compensated by increasing component concentration Component may be filtered-serilized for crucial research

RT = Room Temperature

Appendix 4

Table : Conversion of mg/litre ot μM concentrations for commonly used plant growth regualtors.

Growth regulators	μM							
	NAA	**2,4-D**	**IAA**	**IBA**	**BAP**	**K**	**2iP**	**Zea**
M_l mg/litre	186.2	221.0	175.2	202.2	225.2	215.2	203.2	219.2
0.0001	0.0005	0.0004	0.0005	0.0005	0.0004	0.0005	0.0005	0.0005
0.001	0.005	0.0045	0.006	0.005	0.001	0.005	0.005	0.005
0.005	0.027	0.023	0.028	0.025	0.022	0.023	0.025	0.023
0.01	0.054	0.045	0.057	0.049	0.044	0.046	0.049	0.046
0.05	0.27	0.226	0.285	0.246	0.222	0.232	0.246	0.228
0.10	0.54	0.452	0.570	0.492	0.444	0.465	0.492	0.456
0.25	1.34	1.13	1.43	1.23	1.11	1.16	1.23	1.14
0.50	2.69	2.26	2.85	2.46	2.22	2.32	2.46	2.28
1.0	5.37	4.52	5.71	4.92	4.44	4.65	4.92	4.56
5.0	26.85	22.62	28.54	24.61	22.20	23.23	24.61	22.81
10.0	53.71	45.25	57.08	49.21	44.4	46.47	49.21	45.62
25.0	134.26	113.12	142.69	123.03	111.01	116.17	123.03	114.05
50.0	268.53	226.24	285.39	246.06	222.02	232.34	246.06	228.10

NAA= a-naphthaleneacetic acid; 2,4-D= 2,4-dichlorophenoxyacetic acid; lAA= indole-3acetic acid; 1BA= indole-3-butyric acid; BAP= 6-benzylaminopurine; K= kinetin; 2iP= N-isopentenylaminopurine; Zea= zeatin.

Standards

1. 1kb of ds DNA (sodium salt) = 6.6×10^5 Daltons
2. 1kb of ss DNA (sodium salt) = 3.3×10^3 Daltons
3. 1kb of ss RNA (sodium salt) = 3.4×10^5 Daltons4.

The average MW of a deoxynucleotide base = 324.5 Daltons

Table : Common conversions of nucleic acids and proteins

I. Spectrophotometric conversions

A_{260} unit of ds DNA= 50µg/ml DNA
A_{260} unit of ss DNA= 33µg/ml DNA
A_{260} unit of ss RNA= 40µg/ml
1pmole of pBR 322DNA= 2.8 µg

II. DNA Molar Conversions

1µg pf 1,000bp DNA= 1.52 pmole
1µg of p BR322 DNA= 0.36 pmole
1pmole of 1,000bp DNA= 0.66 µg

III. Protein Molar Conversions

100 pmoles of 100,000 MW protein = 10 µg
acids of coding

100 pmoles of 50,000 MW protein = 5µg

100 pmoles of 10,000 MW protein = 1µg

IV. Protein/DNA Conversions

1kb of DNA= 333 amino

capacity = 3.7×10^4 MW

10,000 MW protein = 270 bp DNA
30,000 MW Protein = 810 bp DNA
50,000 MW protein = 1.35 kb DNA
100,000 MW protein = 2.7 kb DNA

Table : Lambda Markers (size in kb)

Eco Rl	*Hind* III	EcoRl + *Hind* III
21.226	23.130	21.227
7.421	9.416	5.148
5.804	6.557	4.973
5.643	4.361	4.268
4.878	2.322	3.530
3.530	2.027	2.027
	0.564	1.904
	0.125	1.584
		1.375
		0.974
		0.831
		0.564
		0.125

List of Equipments Required for Molecular Biology Laboratory

1. Ultracentrifuge (Beckman, Sorvall, Hitachi)
2. Refrigerated Centrifuge (Beckman, Sorvall, Hitachi)
3. Biorad, PDS 1000 (Biolistic Gun)
4. Water bath & Environment Shakes (New Bronwick labline Kuhner)
5. Transilluminator with Poloroid Camera
6. Transilluminator
7. Ice Flake making machine
8. Spectronic 21
9. Spectrophotometer
10. Deep freezer (-80°C)
11. Gas chromotograph
12. Thermal cycler (PCR)
13. Analytical Balance (Metier, digital)
14. Light Microscope
15. Stereo microscope
16. Electrophores apparatus with power supply (both horizontal and vertical)
17. DNA sequencing apparatus including power supply (LKB. Pharmacia)
18. Microwave oven
19. Scintillation counter (Packard, Beckman, LKB)
20. HPLC (Schimatzu, Backman)
21. Microfuge (Hereaus, Eppendorf
22. Laminar flow
23. pH meter
24. Fraction collector (LKB, Pharmacia)
25. Sonicator
26. Hybridization Oven
27. Vacuum concentrator
28. Lypholizor
29. UPS (3KW)
30. Computer (Pentium with jet printer) IG BHDD, 16MB RAM
31. Ambient Shaker
32. Automatic Micropipetter
33. Hot plates/magnetic stirrer
34. Vacuum Centrifuge
35. Refrigerators
36. Air conditions
37. Autoclave
38. Ordinary centrifuge
39. Nitrogen gas containers
40. Microtitre plates for ELISA and ELISA Reader
41. Photometer
42. Biolistic gun
43. Electroporation equipment
44. Liquid scintillation analyzer
45. Incubators
46. Sequential timers, temperature and light centroller
47. Radiation counter
48. Double distillation water Unit
49. Vortex
50. X-rays cassettes
51. UV protector glasses
52. UV Chamber
53. Image Analyzer Microscope
54. UV Torch